75

SCIENCE AND POLITICS

SCIENCE
AND POLITICS

by
The Rt. Hon.
VISCOUNT HAILSHAM, Q.C.

GREENWOOD PRESS, PUBLISHERS
WESTPORT, CONNECTICUT

Library of Congress Cataloging in Publication Data

Hailsham of Saint Marylebone, Quintin McGarel Hogg,
    Baron, 1907–
    Science and politics.

    Reprint of the ed. published by Faber and Faber,
London.
    1. Science and state. I. Title.
Q125.H26 1974   350'.855   73-16740
ISBN 0-8371-7227-6

© *The Rt. Hon. Viscount Hailsham, Q. C. 1963*

First published in 1963 by Faber and Faber, London

Reprinted with the permission of Faber & Faber Ltd.

Reprinted in 1974 by Greenwood Press,
a division of Williamhouse-Regency Inc.

Library of Congress Catalog Card Number 73-16740

ISBN 0-8371-7227-6

Printed in the United States of America

CONTENTS

———————————★———————————

FOREWORD

————————————— ★ —————————————

It is notoriously foolhardy for a Minister to publish a book
—particularly when it happens to be written around if not
actually about his own subject.

Moreover, quite obviously, if he does make the attempt he
must sedulously avoid either an account of work in progress,
for which he is responsible to Parliament, or an anticipation
of immediate future policy, which has to await clearances
from colleagues or the result of inquiries actually pending.

None the less I am, so far as I know, the first, and possibly
the only Minister for Science (or of Science for that matter)
in the Universe, and the relationship between Government
and science is new and rapidly evolving and exciting wide
and proper interest. Moreover the problems thrown up are
not, in the main, peculiar to democracies but are common
to all industrialized nations. For all these reasons it is possible
that one with my special experience and responsibilities may
have something to say worth consideration.

I also wish to say a word for our own evolved method of
organization, and, indeed, for the philosophy behind the
creation of my own office, which is the result of some forty
years of continuous development. I happen to think we are
entirely on the right lines in creating a portfolio for a Minister

9

whose work is to preside over, and to be responsible in Parliament for a number of semi-independent bodies like the Research Councils and the Atomic Energy Authority, each of which has an independent life and existence of its own. I even believe that the idea is capable of extension beyond the strictly scientific field. The fact that the exact organization of my own office is certainly not more than a link, and not the final one, in the evolutionary chain, does not seem to me to deprive it of importance or even reduce its value.

Obviously, a Minister in the course of his work makes his views public from time to time. In order to prepare this volume I have made fairly generous use of the material I collected for my eighth Fawley lecture delivered at South-ampton University, for my Commemoration address at the Imperial College of Science and Technology, and for a lecture entitled *Staat und Wissenschaft in einer freien Gesellschaft*, delivered to the Arbeitsgemeinschaft für Nordrhein Westfalen in September 1961. I also have made some use of my Rectorial address to Glasgow University in the autumn of 1960, and of an unpublished paper I had intended to deliver in Kansas City in the summer of 1960.

My personal thanks are due to Mr Peter Goldman, C.B.E., for his advice and assistance in sketching the range of my subject and arranging the material; in particular I owe to him the advice to include my excursus into international law and the final essay in metaphysical speculation, which, although closely allied to the subject I have chosen, are perhaps rather the reflections of personal interests and convictions rather than positive contributions to thought about science and politics.

Chapter I

THE ADMINISTRATIVE THEORY

———————————————★———————————————

When modern science began, not many centuries ago, it was primarily a private venture of individual questing minds. It was outside the range of the ordinary educational curricula and still further from the orbit of political discussion. Technology, at least until the first world war, followed in the wake of science. Apart from defence, it was predominantly the preserve of closely guarded commercial secrecy inside private industry. Today, there is no modern Government which does not pride itself on its interest in scientific and technological matters, and few, at least amongst the advanced countries, which do not have some really remarkable achievements to boast about. So complex has the training of the scientist and the engineer become, that a large part of any national educational system has to be occupied with the essential provision of scientific education, both basic and specialized. So costly are the equipment and apparatus needed for research, especially for much of the indispensable fundamental work, that only a national—indeed, in certain cases, an inter-national or even supra-national—Budget can supply it on an adequate scale. This expansion of Governmental provision, this development of the political interest in the scientific, has rested upon the clear demonstration that a nation's power to prosper in peace, survive in war, and command the respect of its neighbours, depends very largely on its degree of scientific and technological advance. All nations, therefore, slave and free, developed and undeveloped, do their best to achieve a high level of scientific and technological investment within their own borders. With many statesmen, perhaps with most, it is a simple question of the survival of their national societies and the growth of their national wealth that they should do so.

These considerations are fairly obvious and need no emphasis. What is not so obvious, and may, therefore, be in more need of argument, is that—at least in its higher reaches —science is not intrinsically a matter that should, or indeed can for long, be pursued solely for the sake of the wealth or the power that its pursuit may bring.

All genuine science has its origin in the intellectual curiosity of the free human spirit, in its creative genius, in its power of insight. It is not something which *can* be dictated to either by theologians or idealogues—or, in the long run, by rulers. It may be that they can destroy it by repression, or even starve it for lack of material facilities. But they cannot by their own act make it live. It is doubtful how long they can even cause it to grow by supplying the means of growth. It is certain that they cannot, simply by supplying the means, dictate the nature or scope of the discoveries that can be made—any more than they can commission a play of the quality of *Hamlet* simply by offering a sufficient monetary reward. Far less, indeed; because it is in the nature of scientific research to discover something which is intrinsic-ally *new*. If anyone were able to predict, or pre-empt, the answers to scientific problems, they would not be genuine scientific problems, for the essence of the matter would already be known in advance. The discoveries of a Newton, an Einstein, a Darwin, a Planck, or a Rutherford are genuinely not known in advance, and, when they are made, their practical applications are not immediately perceived even by the inventors. This is something which lies at the root of the subject, and which statesmen and their people would do well to ponder more deeply than they do at present.

To say all this is not for one instant to lose sight of the spectacular achievements of science under the patronage of Government: nor is it to doubt that these achievements will be multiplied in our time and that a great deal of my own time whilst Minister for Science must be spent precisely in seeking to multiply them. Nevertheless, there is a sense in which I continue to regard the results of Government patronage as something of a paradox, and I am tempted to speculate whether, in the end, the influence of an interest by Government so obviously materially motivated will not obscure the very insights on which creative science is essentially based. In Russia, and perhaps even in America, to look no nearer home, I see some signs that this may ultimately be so; if it does not prove to be the case it will be because the creative human genius in science, as in the arts, may prove too strong for the material motivation of the Governments which seek to direct its course.

However this may be, I begin by insisting that pure science is intrinsically as much a branch of culture as history, philosophy, or poetry, and that the relationship between Government and pure science is essentially the same as that between an enlightened Government and the artist—that is, the relationship of an enlightened patron and not that of an employer.

My argument is that at its most vital and critical growing point, that of fundamental discovery, the Rutherfords and the Darwins have to be treated with not less respect than the

Shakespeares and the Goethes. Historically, science is born of two highly respectable parents, the belief in the uniformity of nature which was the natural child of monotheism in religion, and the freedom of expression and thought which was the product of the Renaissance. I would claim that if Government sought to interfere with the scientist in the absolute freedom of his explorations, and the integrity and independence of his speculation, it would ultimately destroy his real source of vigour, or alternatively frustrate his purpose by undermining his confidence and his will to co-operate.

*   *   *

But to this thesis there is a clear antithesis. The constant temptation on the part of Government to do this thing is matched by a constant need on the part of the scientist, especially in some ways the purest of the pure, to seek Government help on an ever-increasing scale. In most fields the day of the gifted amateur in science is rapidly closing. As I began by pointing out, some of the research tools of the modern scientist cost, almost literally, the earth, whilst the education of the modern scientist is not only protracted and complicated, but almost as expensive as the tools he employs. He is increasingly dependent on the patronage, in some form, of Government.

Yet in exercising its patronage the Government is not, and cannot be, disinterested. The immense advantages which enure to a state that makes use of the discoveries of science—and trains its citizens to make them—are such that all states make some sort of an attempt to do so, and compete eagerly with one another in the attempt. Who will deny that, in the process, both the state, and science, have to some extent been corrupted? Three-quarters of British Government money spent on science comes directly from the defence budget. The proportions in America and Russia are probably greater; indeed, it is questionable whether the basic organization of scientific endeavour in Russia makes any serious attempt to give civil science, as such, an independent life and existence at all. In Britain, the creation of a Minister for Science whose terms of reference excluded defence was something of an act of faith, and has come in for some criticism. It is probably fair to admit that a price had to be paid for this. Nevertheless, I remain obstinately of the opinion that, in the long run, the marriage between science and defence is corrupting, and will at best turn science from a liberating to a destructive force, and at worst ultimately dry up the wells of inventiveness in the scientist himself. Indeed, there is a point at which technology itself will perhaps corrupt and destroy the scientific inventiveness on which it was based.

I, therefore, earnestly hope that at least a proportion of scientists in all countries will retain their freedom and integrity. It may be that the mere necessity to train all scientists in the genuine techniques of free speculation and invention will cause them to break their shackles. Certainly I am glad myself that my own position of Minister for Science is organized in such a way that neither defence nor even the short-term requirements of technology have been allowed to claim complete mastery over scientific endeavour.

But I would think that the strongest guarantee of a correct balance lies in a healthy relationship between Government and Universities—not simply as regards the faculties of natural science, but generally in relation to their whole cultural purpose. There may be many means of making adequate Government support available without infringing academic freedom. We are naturally proud in Britain of our system which allows Government funds to be administered, in effect, by the Universities themselves. I am not saying that this is the only possible way of achieving the desired object. But to achieve this end by one means or another remains, I am convinced, at the basis of any healthy relationship of Government and science.

Indeed, whatever the temptation to militarize or commercialize science, there is a practical consideration which tends in the other direction. No country in the world has ever successfully set up a Department of Science, in the sense of a Ministry directly controlling the pace, the scope and the methods of scientific research. This is due to two considerations. The first is that the strategic planning of science cannot be undertaken without full participation of the scientists themselves—and by these I mean, not simply a staff of administrators with scientific degrees; but men and women who actually carry on scientific work, whether in Universities, Government research stations, or industry. This is as true of socialist Russia as it is of capitalist America, and it is not less true of Britain. Until a short while ago, Russian science was run by the Academy, and though the recent change by which it is now subordinated to a co-ordinating committee is almost certainly a move towards tighter control by the state, it is not a means of taking scientific decisions out of the hands of scientists or reducing science to a mere branch of Government administration.

There is also a theoretical argument which tends in the same direction. There is a sense in which there is no such thing as science, but only sciences. Another way of stating this is to say that science is in fact an all-embracing term, and that scientific researches into particular fields are functions of those fields and not of a comprehensive entity called science. From one point of view, medical research, for example, bears a much closer relation to the climate, population, health, diseases and economic activities of a nation than to their nuclear physics. In terms of science, as distinct from economic policy, it would be meaningless for a Treasury official to try and block a grant for medical research on the ground that the money was needed for a synchrotron. It is true that both projects must take their stand in the queue for the general investment programme. But they are related to other items in the programme more closely that they are related to one another.

What is wanted, therefore, is not a single scientific general staff forming part of a single Department of Science, and settling priorities between the synchrotron and grants for medical research, but a series of scientific general staffs or Research Councils directing scientific research in the general fields to be occupied; independent of day-to-day bread and butter work in those fields, but in close contact with them. At the same time there must be, of course, some body to perform the more generalized function of supervision, looking for gaps or signs of imbalance, and fitting the parts into the general economic, cultural, and social pattern of the entire community. This is the framework of our present British organization. And so, taking medical research again as an example, we have a Medical Research Council, independent alike of the Department of Scientific and Industrial Research and of the Ministry of Health, but under a Minister for Science, and of equivalent status to other Research Councils serving other scientific fields, commanding the respect and enjoying the co-operation of the scientists, but serving the public interest through its subordination to a Minister responsible for science, who, in turn, is advised by a general Council on Scientific Policy.

The question arises, however, whether such an organization does not give rise to an ivory tower mentality on the one hand, and lack of scientific-mindedness on the other. Critics have undoubtedly claimed that this is so, and it is sometimes said that on the one hand British industry is too little conscious of the need for technological progress, and on the other hand that the Research Councils are too strongly academic and too preoccupied with fundamental research. I certainly agree that the organization which I have been attempting to outline is not in any way a substitute for adequate technical or qualified staff at the executive and administrative level in industry or Government; indeed I would go further, and say that it will only work at full efficiency on the assumption that such staff are available and active at every level, and in close contact with the Research Councils. This, however, does not diminish the case for the Research Councils, which is based on the need for detachment and independent supervision, and not on the supposition that they are substitutes for an appreciation of scientific and technical factors in those who are responsible for management in Government or industry.

This leads me to another generalization about the nature of my subject. Science is, or at least it should be, all-pervasive, and will not achieve its best results until this is the case. You can never make industry research-minded, whether privately or publicly owned, simply by providing research in Government institutions or Universities. And, if British experience is any guide, you will not even achieve it only by encouraging co-operative research by associations of individual firms—valuable though such associations be. There is no substitute, at any level, for science *on the ground*, that is, within the individual industrial or commercial firm, within the individual teaching institution, within the individual Government departments and nationalized corporations, and even within the home circle. This may seem impossibly indispensable. Moreover, from every wider point of view, any attempt—perhaps in order to obtain rapid results, perhaps for short-sighted financial reasons, or possibly even for less creditable reasons of political or ideological obscurantism—to generate a race of scientists interested only in their own specialities, to the exclusion of general culture or even of other scientific disciplines, would, I hope, be bound to fail. If it were not to fail, and if it were taken to anywhere near its logical conclusion, it would certainly be the most disastrous enterprise upon which a developed human society could possibly embark. Such a race, narrow, blinkered, unbalanced, humourless, frustrated, and—over a large range of human experience—inarticulate, would be one of the most evil that the human imagination could devise; probably a race of slaves designed to serve a race of still more uncultured masters; and, if not slaves, illiterate but technologically proficient tyrants to destroy or enslave the society which brought them into being.

To this first proposition, however, I would add an immediate footnote. In this sound general education of which I have spoken, science should form an integral part from the first. From the time that a child can read, and count up to ten, an element of scientific knowledge and thought should be injected into the curriculum of what he learns, even if it is only the most generalized statements about the Universe, the microcosm of the atom, the organization of life and evolution, and the nature of chemical compounds. The reason why science is regarded by so many otherwise cultured people as a hidden mystery is because so little of it is injected into general reading and conversation before the age of ten. We should not learn our culture—our history, our religion, our ethics, our physics, our mathematics—in a group of watertight, vertically divided compartments, or in a series of to a large extent, of the formal courses of the University. But this is no longer the case. The men whom I am now discussing take about thirty years to train—five in the nursery, ten in the school, three or four in the University as undergraduates, four or more learning to research, and perhaps another five working under some kind of supervision. Their research apparatus—originally made of glass and brass in an attic or a garage or an outhouse—may now cost millions of pounds like the proton synchrotron at CERN, or the fast breeder reactor at Dounreay; and the supporting structure of constructional engineering, electronics, computers, served by technicians and craftsmen, employs whole industries and forms the basis of life in whole townships and communities. Of those who start off on this road, many fall by the wayside. On the genius of a small proportion out of the very few who survive the race depends the success or failure of all.

Here my first proposition is that at the highest level, it is ruinous to separate research and teaching. It follows that the most advanced scientific work should be located in or near the Universities, where this is possible, or directed by University personnel. The contrary course—the creation of separate Government-sponsored institutes—is attractive and is sometimes rewarded with spectacular results. Moreover, considerations of national security may sometimes dictate it. But in the end it can be a primrose path, as everyone from Russia westwards who has tried it knows. You may gather together all the most promising intellects in a subject, build them a town, equip them with all the most lavish facilities, and the thing will go off like a bomb—at first. But in five years time? In ten? In fifteen? What then? Unless active steps are taken to redress it, there is a danger that intellects will be staler, new blood will not be forthcoming, the stimulus of youth will be absent. From a proportion of those engaged, the research potential may have disappeared; even the scientific questions will have changed. Instead of a research institute you would then have a monastery which had lost its faith— isolated, stereotyped, remorselessly carrying on activities with a diminishing margin of useful results.

I am of course exaggerating a little, to make my point, and I recognize that there have been, of course, remarkable examples of Research Institutes where the dangers I have mentioned have been avoided, it would seem indefinitely. In Britain I would cite the National Institute of Medical Research and the National Physical Laboratory, and on the Continent the Institutes of the Max Planck Society. But if the dangers have been avoided it is precisely because they have been recognized as dangers. It is also because, to use the terminology of the Zuckerman Report, there is a distinction between 'pure basic research' and 'objective basic research' which is stimulated primarily by a specific technological need and calls for a planned approach. But as a general proposition I adhere to the view that there must be the constant stimulus of teaching, or research will tend to die. There must be the constant inter-penetration of the mature intellect with the adolescent, or it may petrify. There must be a constant outlet for research workers who are ready to pass into teaching or administration, or research may even degenerate into a racket.

Therefore, where possible, locate your best research in University or teaching institutions. If you cannot, you must do the next best thing and arrange for constant interchange of personnel, and regular exchange of ideas between them.

This does not mean that there is no place for Government Research Institutions, or institutions of research collectively organized by or for the benefit of private industry. On the contrary, as will be seen, my thesis is that there is room for more and more of these. My point is only this: that however much these bodies proliferate they should where possible locate their best research in or near Universities, and that where they cannot do so they should promote the closest possible links with teaching institutions of all sorts.

To begin, therefore, with the Government. Most Governments—directly or indirectly—build the roads, manage the telephones, dispose of the sewage, construct and maintain the harbours; many manage the railways, generate the electricity and gas. All are concerned with standards of measurement, heat, light, noise, health, water and air pollution. Most are concerned with the nutrition and health of their citizens, and few can divorce themselves from the efficiency of agriculture. All these activities throw up problems of research and development, many of which cannot conveniently be carried on in conjunction with teaching. Some activities are intrinsically public in character. Others can efficiently be financed only by heavy drafts of public money. It will be a very long time before agricultural research is paid for entirely by farmers. Nor can the pharmaceutical industry, the medical profession, or private generosity, finance medical research on a completely adequate scale. Nuclear and high energy physics has too close a relation to defence to be limited to Universities and is too vast and unremunerative a subject to be confined to industry except in its applied forms. Even in the sphere of industrial research proper there are many general questions, like those connected with fatigue and creep in metals, which lend themselves to Government research financed by Government, because they are not so closely connected with any particular private industry as to attract sufficient private finance.

The formal relationship between Government and science in Britain is embodied in an institution which we owe in part to the genius and foresight of the first Lord Haldane, and which has on the whole won the confidence of the scientific world and created the means of providing Government finance without sacrificing either academic independence or scientific integrity. This is the institution of the Research Councils: the Department of Scientific and Industrial Research, the Medical Research Council, the Agricultural Research Council, and the Nature Conservancy.

The thinking behind the creation of these Councils is to separate the activity of research from the executive business of Government and economic activity—keeping medical research separate from the Ministry of Health, for instance—and to concentrate it under a number of Councils, composed mainly of working scientists and industrialists with real executive responsibility for the research programmes under their control, with separate budgets negotiated with the Treasury, and invested with the responsibility, within their several spheres, for financing the training of research workers no less than carrying forward the business of research. All operate stations or institutes of their own. All finance research in the Universities. And one, the Department of Scientific and Industrial Research, co-operates with particular industries in financing joint programmes of research carried out by organizations known as Research Associations, managed primarily by the industries concerned on a co-operative basis, but attracting also Government grants through the Research Council.

It would of course be a logical and attractive arrangement if one could go on to report that the fundamental or basic research in Britain was now carried out in teaching institutions, and that the applied research was all done in Government or industrially-based research institutions. Experience, however, shows that, although this may in theory and in

25

THE ADMINISTRATIVE THEORY

How far, for example, is it possible for *national* Governments and administrations to guide and fertilize science, without warping its character and limiting its scope? I have already pointed out that the gifted amateur, and even the rich private or industrial patron, whether individual or corporate, cannot possibly cover the sweep of modern science and technology, or even the training of the modern scientists; the needs have reached the national scale.

But the nation state also has its limitations, even financially, and more still in its terms of reference. It is surely no coincidence that, in one field after another, science not just on a national but on an international scale is beginning to become an actuality. The International Geophysical Year is only one example, in which even East and West co-operated. Europe has recently begun to plan to enter co-operatively into space research. Britain and America are co-operating in the launching of the first British satellite experiments. Western Europe countries co-operate, under the auspices of the European Nuclear Energy Agency, in the Dragon High Temperature Reactor project at Winfrith Heath, and in the Halden project in Norway. I have already mentioned the internationally owned proton synchotron at CERN. There are projects on foot for both European and Atlantic Universities. Much of the aid to under-developed countries will be both co-operative and technical. What is the future and what are the prospects for enterprises of this kind?

I would say that a very considerable scope exists for international scientific effort. In the long run, I am quite certain that the really ambitious space research, if it is to continue indefinitely, will have to be financed in this way. I cannot myself conceive that either medical or agricultural problems would fail to gain by international treatment. Many of the longer-term projects in engineering, for example, in the realm of thermonuclear fusion or the direct generation of electrical energy from heat, may well come into the same category.

At the same time I would emphasize that international co-operation is no substitute for national excellence. Our experience has been that co-operation is only possible between nations each of which has something comparable to offer to the other. It is neither practicable nor desirable to create international scientific organizations to make up for a basic want of scientific organization or effort in the individual member countries. The new organizations, when they come into being, must be additions to and not substitutions for what is attempted nationally. They will be no excuse for the individual member states to neglect either national education or research on a national scale—although duplication of effort can be avoided, and a scale effect obtained by jointly undertaking projects which are in excess of the capital resources available to any one.

The second question I want to raise lies at the other end of the scale. Hitherto it has been possible for the nation state to stimulate the growth of scientific effort in a variety of different ways—by supporting scientists in the Universities or Research Institutions, and financing their machines and apparatus, and by developing machines capable of practical application in various directions where the State itself is the ultimate user of the machine. But this has meant, in societies both free and servile, predominantly the development of engines of war, even where these have subsequently or concurrently civilian application. Where would the aeroplane, the radio, radar, jet propulsion, or, more recently, space research have been but for the constant urge of Governments to prevail in two world wars—or should I say in three? It is useless for Governments to protest that their

Gagarins and their Glenns have also achieved technical success in realms of peaceful science and technology. Only those nations which have had reason to develop military rockets have actually reached for the stars, and a principal effect of their space flights has been to emphasize the prestige of their military technologies.

Is there no means nowadays of Governments developing or helping to develop new devices which are just as revolutionary but without the stimulus of war or the fear of it? The steam engine in its origin, and the internal combustion engine in its inception, were primarily civilian devices. What must we do now to stimulate development in a purely civilian sense?

I must begin by emphasizing that, as between research and development, it is in the main development which costs the money. In round figures, of the £300 million spent on research and development by Government departments in Britain, about £80 million goes on research and the rest, I estimate, on development, and of this, of course, a proportion that I would not care to guess at is spent on projects which never ultimately reach fruition.

Secondly, I point out that the role of the would-be financier of development is radically different according to whether the financier is, or is not, the ultimate user of the hardware to be produced. One reason why so much successful development has been inspired by defence has not been the over-riding importance of the conception of war in the mentality of modern Government; it has been the simple circumstance that the defence forces as the ultimate users of the hardware have been able to formulate their staff requirements, undertake their feasibility studies, and finally proceed to development contracts with the makers, without the discipline of the necessity to make a commercial project, but with all the enthusiasm of a potential user whose life may depend upon the excellence of the product.

It is not easy, in a free society, to reproduce all these conditions in civilian life—even where, as is by no means usually the case, the Government is the ultimate consumer. In civil aircraft, in railway engineering, in road construction, radio engineering, electricity generation, perhaps in machine tools and computers, and above all in the peaceful uses of atomic energy, some of the conditions exist, and only in the last case have the possibilities been at all adequately grasped—with results which have as yet not been fully seen. Admittedly there are formidable difficulties—the formulation of the requirement, the provision of Government finance for a project which may ultimately enure at least in part to private profit, the placing of a contract selectively (which could give ground for scandal and abuse); but we shall not advance further until we get out feet wet. It is for this reason that we have recently added to the instruments available to the Department of Scientific and Industrial Research this new one of a civil development contract to individual firms. It may be that here too there may be scope for the pooling of resources internationally.

Of one thing I am quite certain. The preservation of a technological lead cannot long be maintained by means of commercial secrecy. Commercial, like military, security always breaks down in the end, and what advantage exists in priority is obtained by industrial experience in production rather than by the maintenance of technical secrets. A nation remains in the forefront technically, not by patenting its devices but by keeping ahead in invention and development, and this is a complex achievement which lies only in part within the province or power of Government to promote. Let us remember that science is an achievement of the human spirit, and resolve that we will foster it for its own sake, not solely for the wealth and power that it will bring. The respect for truth, the freedom of intellects, the moral discipline that it engenders, are worth more than these. Get them for their own sake, and my conviction is that what remains will be added unto us.

Chapter 2

THE EDUCATIONAL BACKGROUND

————————————————— ★ —————————————————

In years gone by, scientific studies succeeded in developing by their own momentum against a good deal of influential opposition from Church, State and University. This led, I suppose, to a certain feeling of tension between the established studies and the sciences. Since the last war, those days have come finally to an end. Science has arrived—both in the public imagination and in the not less important facts of the educational scene. The present proportion of school-leavers, both from the local authority and the independent systems, is pretty evenly divided, but with the weight slightly—and I would say decisively—in favour of science. Future progress is likely to depend upon an increase in the total provision for further education rather than in a nice measurement of what subjects which students must be encouraged to learn. The most important source of untapped talent lies in those who, at whatever level, from the apprentice to the research worker, voluntarily abandon organized education for gainful employment before they have realized their full potential. The numbers of these are still disturbingly high. But they are coming down gradually year by year.

Nevertheless, whilst science, and scientists, no longer have to fight for recognition, there remain many legacies of the old days which are still capable of doing a good deal of damage. Foremost amongst these is the legend of the 'two cultures', or still worse the belief which persists in certain quarters that there is some kind of innate antagonism between the arts and the sciences. I believe that this is a position which must be strenuously contested if either is to flourish to its fullest potential extent. The time is ripe for an alliance between the sciences and the arts. We cannot afford a schizophrenic culture; for the real antagonism is not between arts and science, so much as between the total civilized culture, of which arts and sciences are the warp and the woof, and the absence of civilization, the fanaticism, the prejudice, the passion, the barbarous ideologies, the sheer antinomianism, which threaten to engulf us all. If there is a real division inside the realms of scholarship, it is not between the artist and the scientist, but between the man—scientist or humanist—who views his own speciality in isolation, and the man who regards his specialized knowledge as placing him indeed in a position of vantage, but within the wider context of an undivided world of civilized values and academic standards.

Sir Charles Snow's diagnosis of the 'two cultures' is thus perhaps more a reflection of politics in the Senior Common Room, than of life in the real world. Neither in exegetical nor in fundamental terms does it stand up to examination. For example, are we to classify anthropology as a science or as a humanity? Do we call the decipherment by Michael Ventris of the Linear B. script of the Mycenean civilization as Greek a literary or a scientific achievement? Was the application of the carbon 14 technique to the Dead Sea Scrolls a piece of historical or scientific research? Is the statistical analysis of style in the New Testament manuscript science or divinity?

Or, on a more fundamental level, can a sense of pure intuition be divorced from any genuinely scientific enquiry? Is not the possession of this precisely the factor which in any walk of life distinguishes the genius from the second-rate enquirer?

The truth is that any educated man today is in some degree a specialist, and few of us understand one another's specialities. But that is not to say that there are two cultures, or two hundred cultures. It simply means that, in the highly complex culture of which we are all part, each of us can understand in any detail only small areas of the whole. The doctrine of the 'two cultures' must, therefore, be rejected as a myth. But it is a myth which faithfully reflects the highly artificial structure of our University courses and the over-specialization which the scholarship requirements, particularly of the older Universities, still impose upon the school curricula. It is a myth which could only exist in an academic society where insufficient attention has been paid to the considerable range of subjects which bridge the gap between academic science and academic humanities—geography, economics, management studies, and possibly also the more philosophical aspects of law—and where too little thought has been given to the pruning and modernization of middle-school and sixth-form curricula. It would be no bad thing if in the years immediately ahead we were able to devote rather less time and effort to the problems of bricks and mortar and large glass windows and rather more to what is actually taught.

There was a nineteenth-century zoologist called Haeckel who, in order to explain the appearance in embryos of certain organs no longer used by the complete animal, formulated a doctrine known as the Law of Recapitulation, which stated that an animal during its ontology repeats its phylogeny. According to this law, the embryo recapitulated during its life the entire evolution of any chain through which its ancestors had passed from the protozoon, through the fish, to the creature of dry land, and so on. Although the theory is, I am told, largely discredited and no longer acceptable biologically, there is a horrible sense in which it is almost true educationally. A pupil during his educational life tends uncannily to recapitulate the various stages through which the school curriculum has passed in bygone ages. School teachers tend to perpetuate their own education in that of their pupils. The things they learned at school are after all, the things they know best, and teach most effectively; and the things they teach best are very often the things which are most valued by those they teach. When a new subject appears it tends to be added as an extra. Long after the steamship is invented, it is thought axiomatic that young sailors can best be trained in sail (and for quite a long time this is true); and long after it had ceased to have practical value the teaching methods of Erasmus were perpetuated among the youth of the nineteenth century. Thus the school curriculum tends to become a lumber room, to which things are added, but from which they are seldom taken away. The schoolboy who creeps 'unwillingly to school' becomes a snail in more senses than one. He carries upon his academic back an immense burden consisting of the intellectual home of his ancestors.

The recognition of this is of crucial importance to the teacher of science. It is not simply that our curricula bear the unmistakable imprint of the time when science was simply super-added as an extra. This is true and, I fear, most important. But it is also true that science has been a school subject long enough to acquire a good deal of lumber, of historical intellectual bric-à-brac, of its own. Quite recently,

THE EDUCATIONAL BACKGROUND

I was advised that our own school science curriculum in England could almost certainly be pruned of a significant part (perhaps 20 or 25 per cent) of its detailed content not only without loss but even with advantage—so long, that is, as the living content is retained, and the pruning restricted to dead wood and to unnecessarily detailed factual knowledge. Britain, we know, is a strangely conservative place; but extensive studies on the content of school science courses—particularly in physics—are now in progress in the United States under the auspices of the National Science Foundation, and our own Ministry of Education has now established contact with the Foundation on this very aspect. There can be no doubt that the revision of science (and mathematics) curricula along modern lines is both highly desirable in itself and would also clear the way for all pupils to study science at school in reasonable depth without making undue inroads into other important subjects. Might one also suggest that both Universities and schools have much to learn from a little research into the possibilities of rival teaching methods?

*   *   *

The first thing, of course, is to make up our minds *why* we are teaching science, and for that matter why we want to teach anything.

For the educator, the first reason for teaching science at school is that it provides an efficient and unrivalled medium for the development and exercise of certain faculties which have educational value: the close observation of material things; the ability to think in the context of material things (but not excluding, later on, abstract ideas); the practical application of the rules of logic; the imaginative thought needed to construct a hypothesis; and the discipline involved in testing one's hypothesis by a study of the evidence—and in discarding it if the evidence so determines. In science, the student needs intellectual effort but little experience of life; he is pitting his wits against nature rather than against wiser men; and he is offered the opportunity (or perhaps I might say is subjected to the necessity) of submitting his opinions to the test of experiment. In this way is developed that scientific judgement which is perhaps the most important contribution of science to education, strictly so considered. By scientific judgement I mean the ability to distinguish between charlatanry and good work, between good and bad arguments in the field of material things, and between the purely hypothetical and the well-confirmed scientific theory. Such a judgement is of value at all levels, whether it is applied to the old wives' tale, to the pseudo-science of so many modern advertisements, or to the reputable theories of science itself.

The end product, as the television announcers call it, of education, is not learning but people; and it is as well to establish the point at the outset that we are teaching science not primarily for its vocational or practical value, or even, in the strictest sense, its educational value (great though this is), but because in a scientific age there is literally no individual who can afford not to be aware, at least in some degree, of the sort of thing that science is, and the sort of mental process which the scientist employs. My own belief is that science, and the scientific method, should be one of the pillars upon which education rests. It should be there from the first. It should never be disparaged. It should be an integral part of the whole.

But if the first mistake is to treat science as a comparatively new and essentially practical newcomer to the curriculum, the second is clearly to over-emphasize its completeness and thereby to under-estimate its potential growth and its influence in the future. In all educational studies we tend to view history as a process reaching a climax in the present day. In a lordly kind of way we are apt to divide countries into the developed and the under-developed. But the truth of the mater, of course, is that, in a scientific sense, we are all under-developed countries, and although the rate of discovery has accelerated to such an extent that it is sometimes claimed that nine-tenths of the totality of scientific knowledge is represented by the work of living scientists, it is hardly likely that we have done much more than scratch the surface of the knowledge which the scientific method can achieve. What passes for scientific knowledge today will probably be viewed in fifty years' time as little more than a bundle of naïve and unco-ordinated hypotheses. The most that can be said for it is probably that it represents a true beginning. The young need to be taught science as a young study at its infancy and capable of almost indefinite expansion, not as a corpus of final and unalterable truths.

It is equally important not to overlook the limitations of a purely scientific education as a preparation for the full life of a whole man. If we are to develop fully as individuals we need to comprehend the social as well as the material environment in which we live. Some aspects of scientific training—for example the emphasis on objective judgement and on intellectual honesty—are of profound importance in moral training; and the kind of judgement developed by a scientific training has a vital part to play in our society. Nevertheless it is broadly true that an education in classics, or in the liberal arts, is greatly concerned with the social aspects of life, but little with the physical: whereas a scientific education is concerned greatly with the physical, and little with the social. Present tendencies are to bring these physical aspects more into prominence. Those who specialize in science at school may therefore be less well prepared for life in a social community than those who specialize in classics, which has much more to say about men and women as individuals and humanity as a whole. The problem, which in spite of its intractable nature may be very simply posed, is how to provide the scientist in his training with adequate perception of the social aspects of his environment. This is hardly less important than the companion problem, which is to provide those whose education is primarily classical or humanist with a sufficient understanding of scientific aims and methods.

One further generalization I would venture to make in this context, and that is that there are no 'wonders of science' if the truth were told. What pass for the wonders of science are merely the instances by which science is able to establish that the world is a wonderful place. But that in itself is one of the things which must surely make the study of science—and, indeed, its teaching also—so immensely rewarding. It is not only that every educated person needs to know how wonderful the world is; this is also something which is very good to know—because this is the kind of knowledge that must, in the end, add to happiness.

*   *   *

Let us now consider more exactly what we are aiming at in science education. Here I place three distinct desiderata:

(1) That the democracy, that is, the mass of the people, who, in the West, ultimately decide the pattern of the society in which we live either by their individual choices as consumers or by their collective voice as voters, have enough science to discharge these functions intelligently. This means

that science must be an integral part of everyone's education, that is, of girls' no less than boys'.

(2) That the bureaucracy, by which I mean the people who in every modern state have the power of executive decision both in industry and in Government and society at large (and there are just as many bureaucrats outside Government as in it) have enough science generally to discharge their functions with due regard to the requirements of others, and particularly to know what science can do to aid them in their special sphere of responsibility. This means a highly educated and, where necessary, specialized administrative class.

(3) That the aristocracy, by which in this context I mean those who by their talents and training have the power of making new scientific discoveries and so of keeping the whole fabric alive—either in applied, or in pure science—have the training and equipment to enable them to discharge their task.

To these three desiderata, I add two distinct corollaries to which I attach at least equal importance. These are:

(1) That no scientific education, no matter how specialized, shall fail to recognize both the potentialities and the limitations of science as a moral training, and in particular its limitations as being a part only of the general cultural background in which we live.

(2) That in every scientific education, generalized or specialized, there shall be built in precisely those ingredients which are necessary to keep the balance. This means, amongst other things, specialists in the humanities among the classes which I have described as the bureaucracy and the aristocracy.

Lastly there is a condition of progress which, in Britain at least, is of the first importance—the recognition of an adequate scale and quality of mathematical teaching as a necessary condition of effective scientific education. We are too apt to think of modern science as a purely empirical growth which owes its success to the abolition of *a priori* and dogmatically orthodox reasoning. This is, of course, true, almost a truism, but like other truisms it is capable of being misleading. If not the theology of science, mathematics is at least the Latin. The would-be student, quite early in his progress, comes to a gate through which he must pass if he is to emerge as a fully qualified scientist, armed at all points. On the gate is an inscription in every human tongue: MATHEMATICS SPOKEN. ICI ON PARLE MATHE-MATIQUE. HIER SPRICHT MAN MATHEMATIK.

All this adds up to a formidable educational requirement at all levels and all ages; and I would testify here to my personal belief that we in Britain at least tend to begin science too late, and that it should certainly start in the primary stage (that is to say before the age of eleven), as a general rule and not as at present only in a few schools. Here let me quote some words written a hundred years ago on this very subject by Thomas Henry Huxley, in his Essay on the Study of Zoology. 'It is my firm conviction', he wrote, 'that the only way . . . is to make the elements of physical science an integral part of primary education. I have endeavoured to show you how that may be done for that branch of science which it is my business to pursue; and I can but add, that I should look upon the day when every schoolmaster throughout this land was a centre of genuine, however rudimentary, scientific knowledge, as an epoch in the history of the country.' He would have been glad to see that a movement to develop science in primary schools in Eng-land is at last growing. Children of this age have a natural curiosity about their environment, and with skilful teaching can be helped to make simple enquiries for themselves (following the tradition of scientific method), and can even at this stage gain some elementary understanding of how a scientist begins to work. This is a good preparation for work at secondary schools, after the age of eleven, though it cannot affect the range of work which the secondary school has to cover.

At the secondary stage we have moved very rapidly since the war in developing scientific studies. At the grammar schools an ever-increasing number of children remain to the age of eighteen, and of these an increasing proportion are taking science. At boys' schools this proportion now ranges from one-half to three-quarters of those in the age groups. In other words, the number of our senior and most able boys studying science now exceeds those studying all other subjects. Scientific instruction for pupils less well fitted to benefit from higher academic study though it is developing, is not yet so far developed as at the grammar schools. Yet its importance should not be underestimated. A good deal is known, in some countries at least, of the requirements of industry, teaching, and other services, for scientists and technologists: very little, by contrast, is known of the requirements for technicians. But it may well prove that shortages of technicians are as serious as those of scientists and technologists, and that their education, both at school and later, may be no less important. A major reconstruction of the system of courses for technicians, craftsmen and operatives in our technical colleges was put in hand a year or two ago. This was intended to provide the maximum continuity between education at school and technical college, to adapt the system more closely to the needs of industry, and to reduce the wastage which occurs through the failure of so many students to complete their courses successfully.

THE EDUCATIONAL BACKGROUND

Unhappily there is still an extraordinary difference between the scientific education of boys and the scientific education of girls—and this in spite of the fact that the sexes have for years enjoyed in theory identical opportunities. In all scientific subjects the numbers both of boys and of girls obtaining passes at ordinary level in the General Certificate of Education at the age of sixteen or seventeen, and advanced level in the same Certificate two years later, has sharply increased in recent years. But in physics and in chemistry the boys outnumber the girls many times. It is only in biology that the girls hold their own. A number of reasons have been proposed to explain this phenomenon, and no doubt all have a measure of validity. Vocational openings for girls are less promising; industry is less interested in employing women scientists; there is a shortage of women science teachers; girls are less interested in scientific and mechanical questions (or at least teachers and parents believe this to be so). Of these factors the shortage of women teachers, at least in mathematics, physics, and chemistry, is perhaps the biggest limiting factor, and like so many things of this kind may be self-perpetuating.

This brings me to the crucial question of the supply of teachers. It is almost certain that any serious attempt to clothe with reality the picture of a full range of scientific education, suitably balanced and fortified by an adequate background of mathematics, would reveal in every country a serious shortage of science teachers without whom the project, in its ideal form, can never be better than a pipe dream. Although the number of school teachers in this country has increased in the past decade by about one-third, there are still both in primary and secondary schools, many classes too large for the most effective teaching. Our difficulties have, of course, been aggravated by great increases in the birth rate. In order to solve them we have increased, and are further increasing, the number of places at teacher-training colleges. So we shall be turning out more teachers—and better trained teachers at that, since we are simultaneously extending the period of teacher training from two to three years. We are also pressing on with vigour with recruiting from all sources open to us, including married women, a number of whom return to the service and make a contribution of real value. But although these difficulties apply to teachers of all subjects, they are especially acute in the sciences. Above all we find a shortage of women teachers in physics, chemistry and mathematics, of whom, as I have said, all too few are being trained at our schools and Universities. For University teaching, of course, nothing but qualified graduate teachers will do. In the schools a good proportion of first and second class honours graduates is required to teach the best pupils—at least in the later stages of their course. The burden of the rest of the teaching must be borne by holders of general science degrees, or by non-graduates from teacher-training colleges. If these are to be forthcoming, it will necessarily involve some forbearance on the part of industry in its demands upon trained manpower. I understand that the Soviet educational system is able to draw on a short period of compulsory service in the teaching profession for all graduates, after the fashion of national service. This would not, I conceive, be acceptable in the West, and I would in any case doubt its efficacy in such a form. One would hope, however, that some special arrangements might be possible to meet the needs of industry (which are indeed urgent and incompletely satisfied), while ensuring that these needs do not prevent a sufficient proportion of trained scientists entering the teaching profession. Unless this can be achieved we shall indeed be devouring our seed corn, and we all know what happens to men or to nations who do that.

* * *

My concentration thus far on the problems of teaching must not obscure the vital importance in Universities and colleges of research and training in research. These are important parts of the life of all advanced teachers, and all advanced institutions should have adequate facilities for post-graduate work. These requirements must be safe-guarded, whatever the pressure of the teaching work involved. For the truth is that the two functions—teaching and research—are not independent of one another. Even on the vocational plane, teaching of the order we require for our administrative, our scientific and our industrial leaders cannot be wholly divorced from research—the true pioneering of knowledge of all kinds. Research will tend to become barren, and researchers to die out, if attempts are made to divorce it from teaching. Teaching will become stereotyped and empty, and learning dead, unless it is carried on, in an atmosphere and at a place where the frontiers of learning are being constantly pushed forward, by the people who are doing the pushing.

But how is research to be financed and who is to call its tune? And if the tune is to be called by the spirit of research itself, the λόγος as Plato puts it, carrying us along like the wind, what conceivable motive will be his who pays the piper?

The issue of academic freedom has never been very far below the surface of either academic or political life. At the time of the Reformation, and at the time of the Civil War, the questions were obvious. But in recent years, and certainly until 1939, they were scarcely asked in our own lifetime in

45

Britain. No one grudged the Prime Minister the right to nominate a few Regius Professors; and academic resources depended as much on fees, endowments and donations as upon public funds. But by 1946 it was apparent that the scale of expansion forced upon the Universities by the range and rapid development of modern knowledge, and the insatiable appetite of industrial society for trained graduates, made it necessary that from henceforth the main patron of the Universities would be the State. Happily to hand was the appointed instrument of State and University—the University Grants Committee, brought into being in 1919. I would think that on the whole this has managed to the general satisfaction to administer public generosity without corrupting academic freedom.

In one direction, however, the UGC has required supplementation. Built into University life is an innate conservatism—no bad quality for any institution, but one that needs to be balanced. From their start, the Research Councils—by which I mean the Department of Scientific and Industrial Research, the Medical Research Council, the Agricultural Research Council and the Nature Conservancy—have all encouraged the initiation of new projects of research, and particularly in relation to medical research have founded whole teams of researchers within, if not necessarily as integral parts of, the Universities. In the last four years the total value of the research grants given to Universities each year by the largest of these patrons—the Department of Scientific and Industrial Research—has increased about threefold. Although I do not claim the credit for this, I am rather proud that this has happened entirely within my own periods of office, since I regard the development as wholly to the good.

Since I first assumed Ministerial responsibility for the

Research Councils, University research has been affected by a number of changes—both in scale and in character—in the field of scientific research generally. For one thing, in an increasingly wide range of disciplines the tools of research have become and are becoming machines—and machines each of which is a major engineering project and constitutes a significant item in the capital investment programme as a whole. Apart from the big machines, the accelerators, the radio telescopes, the gas-bubble chambers, there is the problem of the new institutions, sometimes clustering round the facilities of a parent body, like NIRNS (the National Institute for Research in Nuclear Science) round Harwell, yet half resenting and half resented by the parent organization which gave it birth. Closely related to the machines, but potentially larger still, are newly emerging fields of study, too big, almost from the first, for association with individual teaching Universities; too big, almost from the first, to be confined to purely national budgets; often, like space research or the Mohole project, requiring immense vehicles and immense organization for their initiation; sometimes like the European Centre CERN at Geneva, or the Dragon project at Winfrith Heath, involving international funds and wide implications in the field of foreign policy. Already a large number of international organizations, some bilateral, some regional, occasionally world-wide, some for research, others involving development, have sprung into being and require careful finance and clear thinking if they are to develop on right lines.

Clearly all these changes raise important implications for the relationship of Universities to research, of Universities to Government, and of Universities to one another. One such implication that is gradually becoming clear is that in certain branches of science it is impossible for every University to

47

have equal treatment in all spheres from the Government. Ten linear accelerators each costing £1 million will not make up for the absence of one linear accelerator costing £10 million—if one is needed; and it is quite clear that not every University can hope to have a £10 million accelerator. In this country, at any rate for the time being, there can be—or at any rate there is—only one Jodrell Bank, one Nimrod; there can only be a few gas-bubble chambers; and only a few research reactors. Where are these to be placed? Who is to run them? Who is to arrange their research programmes? And, when a number of such various projects are proposed by scientific enthusiasts, how are we to judge which to accept—particularly where these belong to different fields? How are we to correlate the work of NIRNS with the application to DSIR in the field of nuclear and high energy physics—or with the budget of CERN in which we have only a minority vote? How are the differing projects for radio-astronomy to rank between one another or with the applications for big machines in other fields? How is either to rank with space research—another subject in which other countries are involved? Can individual Universities form partnerships or consortia? Or should they all be provided for through a national extramural institution like NIRNS? If the latter, where is the facility to be placed—inconveniently, as some would appear to wish, for all Universities, or conveniently for only one or two? If a facility is given to one University should any—and if so what—conditions be exacted for the benefit of the others? If a facility depends on a particular man, as surprisingly often it does, what happens if he expires or resigns to move elsewhere? All these questions have actually arisen in a concrete form since I took office. Some I have had to decide. Some have been decided, with my approval, by the Research Councils. We are getting a certain amount of case-law and empirical wisdom about it all.

But it is not only a question of big expensive machines that can be located in or near only some of the Universities. Somewhat similar problems arise out of the increasing diversification of subject-matter. How far is it necessary or desirable, for example, to preserve the existing situation in which the scope for post-graduate work by students of a given institution is often limited by the extent of the facilities—either natural or artificial—which that institution happens by tradition or by accident to have in its possession or in its close neighbourhood? Clearly, University statutes were—or should be—made for the needs of learning; and not the needs of learning for the requirements of University statutes. Should University research still be limited in range by the migratory habits of mediaeval monks? Should the availability of freshwater biology as a subject for research in Cambridge or Leeds be dependent on the contents of the Cam—or of the local Yorkshire factory effluent? Should the availability of the Ph.D. to the staff of a Government Research Station depend on the place where each member of it received his first degree? All these questions need public discussion. Government may have views; but in many cases it will be the Vice-Chancellors, the Courts and Senates, the Faculties and the Colleges, who supply the answers. One hopes that they will be the right ones.

*　　*　　*

I come now to a different and even wider issue. What is to be the relationship between the courses of study and research in the Universities and the known requirements of industry? Distinguished academics often write to *The Times* expressing disquiet at the tendency of industry to demand courses from

Universities in fields which are remote from traditional fields and are of deliberate relevance to practical life—for example, management studies. Similar questions are asked in relation to industrial research and in some of the applied social sciences. Is industrial money corrupting, as the author of one letter appeared to think? If so, ought we not to think the same of Government money? With due respect to the University Grants Committee, and to the Research Councils, Government is not self-evidently so disinterested as it pretends to be. Nor perhaps are even academics quite so pure. As I pointed out in the last chapter, three-quarters of Government money going into scientific research goes on the defence budget. More of this than many people think is spent in the Universities—and none of it could be spent at all without researchers.

My own view, for what it is worth, is that in this country we should welcome and encourage industrial money, if only as a desirable and necessary counter-balance to Government money. At least industrial money is not motivated by war or the fear of it. Moreover, there is no academic doctrine more powerful than the motto of the Royal Society: 'Nullius in Verba'. When I am asked, as I sometimes am, whether it is not degrading that people should be asking for private donations towards cancer research, because the Government ought to bear the cost, I reply that I would not altogether sleep easy in my bed if I thought that either I or, better, the Medical Research Council were the only source of wisdom to decide what men and what ideas were worthy of support. The Americans may go a great deal too far academically in their dependence on private munificence; educationally I am quite certain they suffer enormously from the fact that their schools and, to some extent, Universities are in no complete sense a federal Government concern. But I think we may go too far the other way. From my point of view the degree courses available in British Universities are too narrow in scope, and some of them are orientated too academically. I believe the same to be true to some extent of research work.

There is another important aspect to this. One of the most important subjects with which I am called upon to deal is the relative failure of British industry to make use of science and the scientist. The broad truth is that, outside a few main industries, far too few scientists are employed, and if more *were* employed there are some directors who might not—in their present frame of mind—know what to do with them. Not unnaturally, Members of Parliament and others ask me what can be done to improve this situation. Whatever the immediate answer, I am quite certain that the only radical cure lies in the Universities, in a closer contact with industry, both research-wise and education-wise. Contact between intelligent and right-minded people is not necessarily corrupting; almost invariably, it is stimulating; and in my view at any rate Universities have almost as much to learn from contact with industry as industry has to learn from the Universities. Certainly contact is the only way in which either will learn anything from the other. There is a great contrast between the extent of mutual regard and contact between academics and industrialists in, for example, Germany and this country. In my view, neither gain here by the degree of apartheid they practise. And, incidentally, if that apartheid were broken down, both in teaching and research, my own opinion is that the human sciences, which are excessively neglected in this country, would also come into their own—both in the field of post-graduate research and in the field of new schools and faculties.

In Germany there are far fewer obstacles than exist here to the combining of teaching with management or research in industry; and indeed industrial concerns actually operate research institutes of their own inside the walls of technological Universities. I know, of course, that much can be said in support of our own attitude, our ideological division between teaching and production; but need we take it quite as far as we do? When we see the close contact, amounting almost to organic unity, which exists between industry and Universities on the Continent, and when we see some of the results in terms of design, production and sales on the one hand, and vigorous high-grade academic development on the other, we must surely ask ourselves whether the system adopted on the Continent does not also possess certain competitive advantages which we lack. There can be no easy or single way of achieving this end. But one important need is the creation of the kind of climate of opinion in which industry and the Universities turn towards each other quite naturally, and cross the gulf which has too often divided them hitherto.

Thus if the general rule must be that research and University teaching cannot be divorced, the wider development of the subject implies inevitably closer contact between Universities and Government installations of one sort and another, between these and industry, between industries and Universities, between Universities and one another, between Universities and places of research outside the general pattern which cannot conveniently be placed inside teaching institutions, between the scientific body within one nation and others of the same group or even of opposing groups of nations. These are not really rival conceptions for the organization of research; their roles are complementary rather than exclusive, and they are in truth part of a single complex of a constantly developing conception of a scientific life generally.

Chapter 3

THE INDUSTRIAL OPPORTUNITY

———————————————★———————————————

The predicament of Britain in the latter part of the twentieth century is something which concerns us all. It concerns Government, of course. But it is far too big a thing to be handled by Government alone. We must have the intelligent co-operation if not of the whole population, which is perhaps too much to hope, at least of the whole part of the population which in a frivolous and escapist age is capable of taking things seriously. This is the basis of Government by exhortation which comes in for so much ridicule and adverse criticism. But there is nothing wrong with exhortation provided it is clear and to the point. One of the most effective qualities of Montgomery as a General in the field was his ability to take each fighting soldier into his confidence and give him an individual sense of purpose in the over-all plan.

Public life in Britain today is concerned with nothing less than the survival of the nation. In our lifetime, most nations in the world have been over-run by enemies or dominated by dictators. Ours almost alone has remained victorious, and free. Yet of all nations ours is the most vulnerable—in war offering the ideal target for modern weapons of which the nuclear range is only one, in peace surviving on a pinnacle of prosperity precariously poised over a precipice. The material foundations by which the accidents of history and the virtues of our ancestors underpinned our prosperity have gone. No two-power Navy protects our shores; no gold coinage supports prices at a stable level. Other nations are constantly learning the tricks of industrialization which we pioneered. New combinations of peoples obtain the scale effects consequent on union, in markets, in organization, and in political and military power. There is now a danger that Britain may become a cross between a museum and an old curiosity shop —a place where tourists come to see the ceremonies of trooping the colour and the opening of Parliament, and return home carrying away a small case of Scotch whisky, a memento from Stratford-on-Avon, and a few picture postcards of the Tower of London. If this were to happen we should scarcely survive as a people. For we have trodden the road that leads to greatness and back along that path there is no returning. We are still sitting at the big table playing for high stakes, and unfortunately losses in life are not things you can contract out of, like gambling at a Casino.

Of one thing we can be quite certain. The old skills in the old craft industries are not enough. New countries with workers and problems of their own, will learn to make grey cloth and ships and optical glass—yes, and transistors and cameras, and motor cars and bicycles. This island can no longer rely upon the economic superiority of its established technical skills. There is no hope of our retaining competitive power against people with a lower standard of life who are already mastering some of the simplest techniques, and some of the more advanced ones too, of mass production. We shall be successfully competed against, and beaten, in one field after another. We shall be priced successively out of the market in every case where lower wages and equal or almost equal industrial skill can be assembled and brought to bear.

THE INDUSTRIAL OPPORTUNITY

Our recent trouble with the growing textile industries of Hong Kong and the Indian sub-continent, and the great developments of shipyards in Asia and elsewhere, are but examples of what we can expect. Protection will not save us from this competition, because we depend upon our exports for our livelihood and, though you may protect the home, you cannot protect the export market. Nor will we maintain our position by restrictive practices which were designed to save little pools of privileged craft labour in a world where under-employment was the general rule. We will survive by one principle alone—a ruthless competitive efficiency based on an acceptance of the facts of life and an absolute determination always to be one move ahead, in fundamental knowledge, in its application, in choice of materials, in design, in the organization and lay-out of work, in workmanship and in salesmanship. The competition we shall have to meet, therefore, is not only legitimate but intrinsically healthy. The future of this country, its skill and wealth, cannot depend on keeping others ignorant and poor. Our aim must be to ensure that by the time the new countries are beginning to acquire an existing technique, we ourselves shall be moving over rapidly to another. Indeed, so soon as a thing is invented we must share our knowledge and skill in that thing with others, lest we come to depend on that thing, and not on our continued inventiveness for our continued survival.

In creating an up-to-date, efficient and competitive industry, research has an indispensable role to play. But research is by no means a panacea, and it must not be confined to research into the purely scientific aspects of a problem. The scientific attitude is not confined to science, and even in its scientific aspects, the scientific attitude of mind is as concerned to apply knowledge as to gain it. Indeed there is a danger that research can be used as a bromide. If we are unwilling to face facts or to assume responsibilities it is easy, but almost invariably fatal, to call for more research and more education. Research, like the Almighty, has the habit of helping those most who help themselves.

The recent survey undertaken by the Federation of British Industries revealed a pattern of science in industry which is by no means even and by no means reassuring. Only a handful of industries are really science-based at all; some of the biggest industries are among the least. What do I mean by being science-based? It is not simply a question of research. A science-based industry probably does spend a good deal on research and development by highly qualified scientists and engineers. But to some extent this will depend on the nature of the product. Being science-based means a great deal more than that. A science-based industry is likely to have a modern, purpose-designed set of buildings. Its production, design and sales techniques will be modern in outlook. So will its administration, its management, its industrial relations, its industrial medicine. It is not enough to buy a scientific mind. You must have it yourself if you wish to use it in others. We are moving out of the pre-scientific era, and no one, not a managing director, not a Cabinet Minister, not junior counsel conducting a running down case, can afford to have cobwebs in his mind, or to preserve a lot of useless lumber in his mental attics. The training of technicians and craftsmen is as important as the employment of scientists. We can no more afford to have a system of apprenticeship which involves unfortunate lads fetching endless cups of tea for five years for 'Old Fred'—who certainly can teach them no more than he learned himself fifty years ago—than we can afford to ignore the latest scientific discoveries in electronics or nuclear physics. Indeed of the two, I am not sure that 'Old

Fred' is not a bigger barrier to progress than his counterpart in the managing director's chair. During the war we trained the pilot of a Spitfire in a matter of months. Does it really require five years to make a bricklayer, or a carpenter, or a fitter?

What can we do to make our industries more science-based than they are? I want, as I said in the previous chapter, to see a much closer connection between industry and the technological faculties in the Universities and the colleges of advanced technology. This may involve industry in expense. Industry must be prepared to let some of its bright young men return to do post-graduate work after they have acquired experience. It must be ready to lend some of its technical top brass as visiting lecturers. But if so, it will involve the Universities in considerable change of outlook before they accept this and receive the lecturers or better still appoint them as Professors. There is a lot of the 'Old Fred' mentality in the Universities too, and that even in the scientific and technological chairs. But where there are brains in the Universities—and by and large the best brains in the country are still concentrated there—I want more people in industry to consult them professionally. This is one of the things which science-based industries do, and which could be done with great advantage to the country by more industries, particularly in the engineering field. Only so will the academic mind be galvanized into creative thinking about practical, industrial and technological problems and the industrialist take advantage of the best of new knowledge.

I want larger industry itself to assume a great deal more responsibility for spreading the gospel of science and technology to the smaller firms. The Research Associations could benefit the smaller firms even more than the larger firms and many do, but by and large it is the larger firms which support the Research Associations better than the smaller firms. I want to see information centres supported by industry in several of the larger centres. If they do not in the next year or two, they will have to be closed down for the want of the pitifully small sum of £50,000. I want to see the permanent staffs of great Trade Unions following the American lead and keeping highly-paid research workers to maintain a watchful eye to see that the employers devote enough of their resources to research and development. In America they argue—and to my mind quite rightly—that it is upon such use of resources that future increases in wages very largely depend. I should be a great deal less depressed about the future of British industry were I to read that the shop stewards in a motor-car factory had called an unofficial strike because too little was being spent on research and development, than I am when I read about tea-breaks and demarcation disputes. There could be no quicker way of stimulating management out of what may be sometimes its complacency than for workers to take the lead in demanding a more advanced technology. Above all, I want to see the top brass in the management of industry hammering on the doors of the Department of Scientific and Industrial Research and the Minister for Science—bubbling over with ideas for industrial research, demanding more and more work done, creating closer and closer links with the headquarter staff, with directors and stations, offering more and more ideas for development contracts, research in the Universities, the formation of research associations, sponsored research, and even projects for international co-operation in science, which looks like being one of the great features of the coming half century.

I want to see this country an up-to-date country—up-to-date in Government machinery, in education and training, in industrial management, in labour relations, in Union organization, in professional institutions. I want it to stop looking backwards, and look forwards. I want to see it drop pre-scientific illusions in politics and social ethics. I want to see it stop deluding itself with bromides about the experts always being wrong or everything being the fault of the Government, or at least of someone else. I want to see and hear fewer television programmes about the remote past and fewer Westerns about the more recent, and foreign, past. I want to see fewer advertisements of harmful products, and none at all romanticizing nonsense, or suggesting nonsense about romance. In short, I want this country to be worthy of its past, not perpetually trying to recapture it. A little healthy iconoclasm is as good a prescription for survival as respect for tradition, and a profound sense of dissatisfaction is as good a recipe as any—if it is coupled with creative ability—for efficiency in industry or in Government.

*　　*　　*

At the present time we are far, lamentably far, from the ideals I have stated. But we are much further ahead than we sometimes think. We are incomparably further ahead in science than other countries, with the doubtful exceptions of America and Russia the size of whose programmes has enabled them to achieve a scale effect of which we are not capable. But our best quality is equal to their best, and we probably get superior value for money. No one should think that the Europeans have yet matched us in this field, though there is a very great deal we can and should learn from them—not least, as I shall stress later on, in general engineering and engineering design. Certainly it can do no harm now and again to remind ourselves of our country's successes and achievements, as well as our problems and difficulties, as we move forward towards the unknown world of the twenty-first century.

Britain today has the largest nuclear power programme of any country in the world and is the world's largest exporter of radio-isotopes. Half the world's ocean shipping is equipped with British radar and more than half the world's gas-turbined aircraft are powered by British engines. The pencil beam of Jodrell Bank and the Melrose heart-lung machine are both proof and symbol of continued British leadership in radio astronomy and in medical research. Since the end of the war, twenty British scientists have received Nobel Awards —a total exceeded only by the United States—and Britain is the only country to have held the 'Triple Crown', the world speed records for air, land and water. These are assuredly not grounds for complacency or insular pride, but neither do they afford excuse to those political smart-alecks who seem to spend so much of their time cracking up other countries and running down their own.

From time to time I have felt it my duty as Minister for Science to say slightly disagreeable things about the state of research and development in British industry—and in all probability I shall continue to say them. But I should like to put it on record that, were I the Minister for Science in the Soviet Union (where, incidentally, they have not got one), I should be saying far more disagreeable things about Soviet agriculture and agricultural machines, Soviet biology and medicine, Soviet consumer goods and light engineering products—all of which, by and large, are far behind our own in terms of scientific backing and scientific production, and therefore in terms of quality, design and efficiency. I should also have some fairly rough things to say if I were Minister for Science in the United States, Germany, France, Italy or

Japan. I do not believe in selling Britain short in the technological any more than in the economic or diplomatic fields.
Nevertheless, I am bound to complain when I feel we are not
good enough; and there are several fields in which we are not.

The first thing I did when I was appointed Minister for
Science in the autumn of 1959 was to ask the Advisory
Council on Scientific Policy to undertake an authoritative
review of the whole state of our scientific effort in the civil
field, and particularly its balance and completeness. I had
earlier been responsible for civil science (as Lord President of
the Council) since 1957, and when the Advisory Council
reported in 1960 I felt that it was moderately encouraging to
be told that in three years total expenditure, in constant-
value terms, had increased by about 40 per cent; that the
balance of effort, which was hitherto overwhelmingly in
favour of defence science, was gradually beginning to be
redressed; and that despite the big expansion of Government-
financed science, private industry was showing so much more
interest as to be carrying a larger proportion of the total
figure. I was also gratified to discover—whilst recognizing
the dangers of international statistical comparisons—that the
proportion of the gross national product being spent in this
country on civil research and development was not much less
than in the United States. The Advisory Council pointed out
that we could not hope to be in the first flight in every field
of science at one and the same time. That indeed would be
impossible for any nation, however wealthy or whatever the
resources it might have at its disposal. But there were some
respects in which the Council regarded it as important or
urgent on national grounds to build up a greater effort than
that which we were currently showing. Some of these—like
oceanography and taxonomy—were in the fields of pure
science, but the greatest interest was aroused by what the

attitude towards the sciences on the part of (say) a very highly developed classical specialist. I am sure that in this matter there is need for a revised approach.

I have often regretted myself that the word 'engineer' has so many different meanings, implying so many different kinds of mind and so vastly different degrees of skill: the sapper in the Army, the member of the AEU, the man who drives an engine, the Professor of chemical or aeronautical engineering in a great University, or a practising engineer of such incredible diversity as Brunel.

Be that as it may, engineering is surely one of the most exciting professions in the modern world, as the engineer is one of the most crucial figures in the perpetual struggle upwards of the human race. I remember one of the most moving inscriptions in the later Roman Empire came from a military engineer whose task it had been to bring the water supply to a town by means of a tunnel through a mountain. The tunnel is still there. Today, what infinite opportunities open out in this country and overseas to redevelop what is obsolete and provide development where it is lacking. How many of our buildings and large structures of all kinds—I am not referring to those of historic interest or artistic merit—would have been pulled down in the 1920's had they been sited on the other side of the Atlantic? If we could accelerate the movement of London Transport buses through the streets of the Metropolis by about two miles an hour we would save something like £2 million in public transport costs alone every year—not to speak of the saving to countless firms and private individuals. Who is it could save this money? The traffic engineer. Who can save more lives on the road: the policeman, or the engineer who designs a motorway? Who does more for an under-developed country: the maker of speeches, or the designer of a dam? Who but the chemical engineer can make the nuclear fuel for a power station out of ore, or recover on an adequate scale the rare isotope from the common element?

What surprises me most is our inability to glamourize this subject. Whether it be fact, or fiction for the inspiration of youth, the wireless and the television focus their ideas largely on the past, or on the already over-exploited possibilities of inter-planetary travel—which, whatever its potentialities, will never be the occupation of many. The future will lie with the new-style Brunels at least as much as with the Jeff Hawkes and Dan Dares. And as for the Westerns or historical romances, these artless mass-produced monstrosities only damage the minds that attend to them. We have in this modern age an almost infinite capacity for ignoring the obvious. Engineering is by far the most artistic of the sciences. It is not simply that the really classic designs of the past—Brunel's bridges and railways for instance—are essentially beautiful. It is that all engineering design is, in its inmost being, creative. Incidentally, I am not at all sure that it is not a failure to recognize this fact which makes some engineering teaching a little uninspired. It is not for me to dictate the content of curricula. But could not the teaching institutions here do a little more about teaching engineering design? And could not more undergraduate courses include at least one project?

Here, then, is a series of objectives that we ought to set ourselves for the immediate future. Let us bring about a situation where engineering will be recognized in this country—and particularly in the Universities and in industry —for what it is: one of the most worthwhile and constructive occupations that a man, or a woman for that matter, can follow. Let us see to it that the content of our engineering courses is fitted for the requirements of the life that their

THE INDUSTRIAL OPPORTUNITY

Many of the DSIR's fifteen research stations, apart from helping Government Departments and publicly owned industries and services, also do a great deal of work that is of the first importance to private industry. The tower cranes, for instance, which have now become an established and familiar feature of our landscape, have been introduced into this country as a direct result of the work of the Building Research Station. A new hydrostatic transmission is being developed in the National Engineering Laboratory, so versatile that its principle may be used equally to provide drive to a submarine and to a machine tool. The National Engineering Laboratory, in conjunction with the National Physical Laboratory and private industry, has also carried out extremely promising work on the development of machine tool control with the use of diffraction gratings. The Forest Products Research Laboratory has developed kiln-seasoning techniques which can enormously reduce the amount of stock needed to be kept idle by timber-using industries, and can at the same time speed up the rate of factory production. The Torry Research Station has been doing research work on kipper smoking, cod freezing, and the development of deep-freeze on trawlers, which could benefit the fishing industry and the consumer generally. The new ship tank near Teddington, which was inaugurated by the Duke of Edinburgh, has enabled hull design to be studied and improved in ways which can save millions to the shipbuilding and ship-owning interests. The National Chemical Laboratory has been carrying on studies of extraction processes which have made it possible, in the application of the uranium ores, to keep down the price of nuclear electricity. I could go on enumerating work of this kind. I have selected only a few items of special interest to private industry which I happen to have inspected in my term of office; and I have not mentioned the very notable economies and advances which the work of the stations has helped to secure in the public sector—for instance, in the design of roads, bridges and school buildings.

But the work of the DSIR's research stations does not limit the role of private industry. The truth is that the functions of industry and Government in this field are complementary and neither is really doing enough. No industry of any size can afford to do without research, and although some are so large that individual firms can do it all themselves, and some (like agriculture) are split into such small units that it is all done by Government, yet by and large no industry is complete in this country unless it is serviced by co-operative research. I would say that in general a research association is as much a part of the necessary equipment of modernized British industry as a technical press or facilities for technical education.

There are now 53 grant-aided research organizations, supported by about 22,000 firms and with a total expenditure of nearly £9 million. The Government's policy has been, and is, to foster their growth by contributing to their expenditure. But as these institutions develop and mature it is surely right for the proportion of the cost borne by industry itself to rise in relation to the Government grant. This is not because Government grudges a penny of the money spent on research, but partly because it is only reasonable to expect industry to pay for what increases its profits, and partly because in our system of society it is thoroughly undesirable for Government to do what is essentially industry's job. In the decade from 1951 to 1961 the proportion of Government grant to industrial contribution fell from 1:1.65 to 1:2.60 and the amount of industrial money going to the research institutions doubled. Nevertheless, the Industrial Grant Committee of the DSIR has expressed concern—which I vehemently share—about the financial position of research associations serving smaller industries, some of which experience difficulty in raising from the constituent firms a revenue sufficient for their needs. Even if firms are too small to conduct research, as many of them are, they should not be too small to pay modest subscriptions—ranking, incidentally, for income tax relief—to their research or trade associations. And they are never too small to seek out and apply to their own firms the results of other people's experience in research, which is available to them in trade papers and scientific publications, through the information services of the DSIR, and via the now fully operational National Lending Library for Science and Technology.

A complaint often laid against this country, however, is that the successful conclusion of scientific research work is not always followed by its application to industry. Hence, it is argued, we tend to lose our ideas (and our expert manpower) to other countries who have the enterprise and resources to exploit them. Whatever the truth of these allegations, which are often exaggerated, it is certainly true that major technological developments have become progressively more expensive in recent years, and some may be beyond the reach even of the larger firms. Nowadays important new products may take upwards of seven years from the drawing-board to the production-line—and more than ten before the money invested is coming back. What private industries can afford to provide finance on these terms? Experience has shown that from the fields where Government traditionally pays for development, notably weapons, extraordinary results are seen to flow out to other fields far removed from those for which the development was originally intended. Aircraft, radar, atomic energy, and indeed most of the really startling industrial advances of the twentieth century, have owed the speed of their development (if nothing more) to Government finance borne on the defence budget. The question to be asked, therefore, is whether, in what circumstances, or in what directions, science and technology would benefit from similar injections of Government development money for purposes that are not warlike in origin. It is obvious that peculiar difficulties obtrude themselves where the Government are not the ultimate users of what is produced and where the manufacturers are ordinary commercial firms operating for profit.

The National Research Development Corporation, for which the Board of Trade is responsible, was set up in 1949 with the object of investigating new inventions that were not being taken up and helping to exploit those which showed technical and economic promise. Now, in the last year or two, the DSIR has begun to place civil development contracts to complement the work of the NRDC in support of science-based development in industry, particularly in the machine tool field. The industrial recipients of development contracts will contribute substantially to the cost, and a return on the Government contribution—usually by means of a levy on subsequent commercial sales—will normally be sought. If these projects succeed, we may have achieved a major break-through in the relationship between Government and industry in the encouragement of science.

*　　　*　　　*

The relations between Government and science are therefore still in a formative stage. The situation has changed considerably since I first became Lord President, and I do not think it will become stable for many years to come. I do not pretend to have solved, even in theory, all the major problems that are to be solved in assisting the progress of scientific research or speeding its application to our economic and industrial life. I do believe, however, that we have some of the principles in our hands and that, whilst we may not know the detail of the road, we can be fairly certain that we are moving in the right direction. Sometimes I read, in published prints by progressive persons, attractive prospectuses for what is called 'scientific planning'—with sometimes a science budget or a scientific general staff added, or even a demand for a 'real' Minister of Science, with 'real' powers. Is there anything in these ideas? Or do they in fact add up to the kind of charlatanry which too often disfigures the public life of the modern world?

The first thing to realize is that, from the point of view of planning, science cannot be isolated from politics or economics. There may or may not be a case for a national economic plan; indeed, much current political argument centres round the senses in which such a plan, or plans, can be made for the nation or for individual industries. In such a case the scientist and the technologist would, or at least should, have to play an important part in the planning. But what is plain is that, except in the most abstract realms of pure science, there cannot be a scientific plan which differs from the economic plan or can even be isolated from it. There is no set of circumstances in which the oceanographer, the nuclear engineer and the lepidopterist meet together to discuss a common national plan which has meaning except in the context of politics or economics. Not only can science in general not be divorced from economics or politics, but the individual sciences *can*, to some extent, be divorced from one another. They may each cross-fertilize the other; equally they all form part of a vast single corpus of knowledge; but this corpus of knowledge is not purely scientific, and the

70

relationship between the activity of one branch of science—let us say the research of a man investigating the cancer cells of mice—and that of another—say, looking at the transmission of a motor car—is not necessarily any closer than the relationship between either of these activities and the great complex of medical, industrial and engineering problems in which in each case the research is directly related. That is why in this country we do not have a single scientific general staff, but, if you like, four.

The second thing to insist upon is that, whilst a degree of Government control is the inescapable concomitant of Government finance, any such control, even in principle, is immensely difficult to organize in the scientific field. Partly because of its technical character, but even more because it is the product of the free-ranging quality of the human intellect, science is intrinsically something which cannot be bought or ordered about. Hence the two great weapons of modern Government, finance and administrative direction, cannot be applied in the ordinary way. In particular, money is neither the customary guarantee of progress nor the customary instrument for controlling extravagance. This may be heartbreaking alike to the Treasury official and to the popular enthusiast who reckons that 'we do not spend enough on research', whether in industry or elsewhere. But it is one of the main facts of life to those who are responsible for the relations between science and Government. No one can command success just by giving orders and spending money. The best one can do in commissioning research is to support first-class people and promising ideas. The best one can do for economy is to ask the experts to refuse support to the bogus or the second-rate. If there is a dearth of ideas or of people of adequate quality available and willing, it is better to spend one's money on something else—probably on education, as the best means of supplying the deficiency of men and ideas. But, conversely, even when one has spent money on first-class men and promising ideas, one has to reconcile oneself to the possibility of failure—which in any given case is at least as large as the prospect of success—and in any event to the lapse of a long period of time during which no one will know whether the Snark one is pursuing is a veritable Boojum or not.

The third thing I am most anxious to get across—and which so far I have failed to get across—is that the tidy, ambitious and grandiose schemes which I am constantly being invited to adopt for the aggrandisement of my office, and the enhancement of my personal reputation, would in effect be reactionary and restrictive of my true activities. In Government it is important not so much that there should be a Science Minister as that all Ministries should learn to regard the application of science, each in its own particular sphere, as one of its main responsibilities. Education, Power, Transport, the Post Office, the Defence Ministries, Housing, Agriculture—all such Departments must have scientific staff of some kind as part of their organization if they are to function at full effectiveness. Research must be carried on, and applied, by the nationalized industries and private firms. There is not, and there cannot be, a central organization to carry all this out; the function of any central organization must be to stimulate and encourage this work in others, to fill in gaps, to conduct generalized researches not apt to the functions of narrower institutions, to create organizations and institutions where these are lacking. The function, as I said earlier, is not that of an employer but of a patron; in a sense, indeed, it is that of an impresario.

Finally, there is one political question I wish to dispose of, not for personal reasons, but simply in the interests of sound thinking about good government. Science is, of course, in its very nature, technical. A good many people, in viewing its needs in relation to government, have allowed themselves to be unduly oppressed by the fact that those who have to take important decisions in Government must often be confronted with technical questions of great importance which neither their general education nor their professional training has qualified them to answer. This is an important issue. But it is not (as a glance at the lyrics of *H.M.S. Pinafore* will remind us) an issue confined to science—or even to Parliamentary democracy. Whether we are governed by a Krushchev (who is said not to have been able to read till he was twenty-three) or a Kennedy, a Mao Tse-tung or a Macmillan, an Adenauer or an Azikiwe, there will be many questions which Ministers who are not qualified experts in the field cannot answer from their own experience. Moreover, even within the field of the natural sciences, specialization is now so great that it is doubtful whether a single man can be found who can speak with authority on more than a limited part of it. So we need not be unduly depressed by the fact that we are governed by amateurs, nor waste our time searching for a scientific genius to supplant them, or even for a Lindemann or a Tizard to be their grey eminence. What we need is an administrative machine with a Minister at the head, capable collectively of isolating the right questions, conducting the proper discussion of them with the right people inside and outside the Government machine, rendering intelligible the real arguments, providing at various levels of authority the means of arriving at rational conclusions with the human material in fact available, and finally supplying suitable means for carrying these conclusions into execution.

It is not the amateurish quality of rulers which raises the real problems in the relationship of science and Govern-

Chapter 4

THE INTERNATIONAL CHALLENGE

———————————————— ★ ————————————————

The age in which we are living has been one of rapid, revolutionary and cumulative change, affecting every aspect of human existence, shaking every human institution, unsettling every system of human thought. There have been centuries when the changes due to the inherent instability of human society have illustrated, rather than upset, the fundamental assumptions of ordered life. The transition from the fifth to the twenty-fifth dynasty of ancient Egypt made no difference to the irrigation system of the Egyptian farmer, to the local boundaries he measured, to the crops he sowed, to the festivals he celebrated, to the gods he worshipped, to the recurring miracle of the Nile flood by which he lived. A greater gulf separates us from the world of Dreyfus and Oscar Wilde than separated Akhnaton from Cleopatra—or even Julius Caesar from the Duke of Wellington. A network of Roman roads lasted well into the twentieth century; many a train still trundles through the cuttings and over the embankments engineered by the Irish navvies of the early Victorian era; and, as often as not, is powered by a locomotive not so far removed from Stephenson's Rocket as we should like to think. But, for all that, the internal combustion engine, the aeroplane, the discovery of radio waves and

75

nuclear fission, the thermionic valve and the transistor, the computer, the rocket motor and who knows what in biology, biophysics and biochemistry, have set changes in train in a single human lifetime which cannot stop short at the physical. The social, the political, the moral, the aesthetic, the religious assumptions on which human life is built are shaken to the foundations. The political map of the world is being redrawn; and we in our small island have, like Job, escaped only by the skin of our teeth from catastrophes which have overwhelmed neighbouring societies seemingly more powerful.

It is easy, but necessary, to remind ourselves that these convulsive disturbances have a material origin—the application of scientific means to every department of human endeavour, including manufacture, transport and the communication of ideas. It is not so easy, but no less necessary, to escape from the illusion of normalcy—a present we can retain, a past to which we can return, a future end beyond which change will cease to operate. In our lifetime, or that of our children, these things will never exist. The changes in the next fifty years will be not less catastrophic than in the last fifty, and no less convulsive in their impact on political and social institutions.

Our survival into the twenty-first century will depend on our adapting our institutions to these changes. For the great challenge of our time is presented by the fact that the constitutional machinery of the world has proved inadequate alike to its material needs and its spiritual aspirations. This is an explanation of the paradoxes amid which we live. We live in an age of material wealth more abundant than that of any period in the history of the planet. Yet this wealth serves only to set off the need and squalor in which the majority of men and women still exist, and to excite human cupidity without the immediate possibility of satisfying human aspirations. We live in an age of great military strength. Never before has our own country, for example, disposed of armed forces of such terrifying power. Yet the age of greatest military strength has proved perhaps the period of greatest insecurity, since never before have Britons lived within minutes of destruction. We live in an age of rampant and irresistible nationalism. Never has there been a time when men and women have been more conscious of race, or more unwilling to be ruled by people of another tongue, colour, culture or religion. Yet one abiding lesson of our time is of the failure of the nation state in isolation to provide for its own people either peace, or prosperity, or even human dignity. Another is the certainty with which an unbridled nationalism leads to war. From all this it is clear that the existing categories are insufficient. There is no contemporary human society whose needs—economic, social, political, even military—do not transcend its national boundaries.

The East has an answer to this; but its answer, the answer of Communism, is a world-wide conspiracy against the human race, designed to reduce it to the subjection of a self-confessed dictatorship dominated by the caucus of a single Party. Has the West an answer? Can the West produce a political idea less offensive than imperialism, less anarchic than nationalism, but no less positive, no less dynamic, no less successful? The tragedy of our ideas and ideals to date is not that they have failed, but that the limitations of the political institutions under which they have existed has prevented them from being tried on a scale adequate to ensure their success. We have not been thinking big enough. We have been too concerned each with his own backyard. We need an economic, scientific and cultural unity, and a political apparatus of consultation and co-operation, far closer and more sophisticated than we have so far known, which at the same time can form the exemplar and the nucleus of a world community when the time has come for other nations to wish to join. Can we build this in time? Can we create a state of society so good over so large a part of the inhabited world that neither the uncommitted, nor the committed in a contrary sense, dare call in question the extent and glory of our achievement? It is upon the answers to these questions that our survival is going to depend.

*   *   *

We must realize at the outset that in this battle of ideas in which we are engaged, science and technology are benevolent neutrals, willing to ally themselves with either side which seeks their aid. Have we in our Western society sought their aid enough? It seems to me that in the criticisms we make of the Communist States our political and religious leaders attach too much importance to the barbarous nonsense of their Marxist ideology, which, being false, can only be a source of weakness to them, and give too little attention to their wholehearted pursuit of scientific and technological aims, which, being practical, are far more likely to do serious harm to us. Nor is it sensible at all to suppose that sound political or religious doctrine will do instead of science and technology. Indeed, no political or religious system which does not inculcate logical thinking and self-help, which are the essence of the matter, is fully sound even within its own sphere.

I do not, of course, under-estimate the natural affinity between science and the free or open society. One often hears it said that science and technology flourish best under dictatorships—or even that some scientists tend to favour ruthless and authoritarian governments. I do not believe this is true. Hitler ruined German science and technology for a generation—and may even have lost the war when it came —by his persecution of Jews and moderate-minded men who happened to be scientists. On the whole scientists tend to flourish best in a climate of free opinion. Our country's original lead in science and technology was in itself not unconnected with out liberal institutions. Science is usually scornful of *a priori* authority such as is claimed by all forms of political orthodoxy, and essentially critical of theories which do not stand up to experience or which cannot survive the cool appraisal of reasoning criticism. It may not always be favourable to an atmosphere of religious enthusiasm, but political fanaticism is, I should have said, far more fundamentally and irreconcilably opposed to the spirit of science than is religion. Certainly whoever retains the scientist in the employment of his society or seeks to breed a race of scientists in his schools and Universities, though they may for a time neglect the humanities or despise the ordered teaching of morality and virtue, will in the end create and harbour an army of men and women rebellious against the fatuous assumptions of a political creed which is unbased on human experience and unwilling to accept as part of its natural machinery the self-adjusting mechanism of rational discussion and argument. In the long term these are grounds for hope on the part of the West.

Nevertheless we have to recognize certain dangers inherent in democracy. No democracy has so far shown itself willing to spend a percentage of its national income on formal education equal to that of Soviet Russia, and none has been prepared so thoroughly to reorganize its curriculum in the interest of mathematical and scientific teaching. Unless we act vigorously and in time, we shall be paying an increasing price for these omissions as the years roll on. Moreover, where science thrives in democracy, it is of necessity deployed over a wide front and not switched and concentrated to anything like the extent possible under an authoritarian régime. This concentration of effort cannot be fully initiated by a democracy in time of peace. But the fundamental achievement of the Soviet economy, which is the extremely successful marriage between science on the one hand and engineering and technology on the other, can certainly be reproduced by the private enterprise or mixed economies of the West if they are really prepared to bestir themselves in pursuit of this objective.

It would be a mistake to under-estimate the magnitude of the challenge or the absolute necessity of facing it together. We must not think that the already narrowing margin of scientific and technological superiority enjoyed by the West will cease to narrow. On the contrary, unless we pursue policies altogether different from those which I think we are following now, the gap will narrow still further and still faster. No doubt America, Britain and Europe are, on balance, still well in the lead—despite the carefully stage-managed demonstrations of space men and ships and the rest. I would expect us still to be well in the lead in 1970. But when I compare *our* relatively wasteful use of our vast resources and national income, and *their* deliberately thrifty, carefully deployed use of their inferior resources; when I compare the four per cent or even five per cent of our national income spent on public education with the seven or eight per cent of theirs; when I reflect that, of that figure, the greater part is concentrated probably on the production of mathematicians, scientists and technologists: then I feel bound to warn that our material advantage is dwindling away, that the rate at which it is dwindling is increasing, and that in some fields we are already behind.

Moreover, if the gap between ourselves and the Communists is decreasing, the gap between ourselves and the undeveloped countries is stable or increasing further. This alone creates a situation of great danger. It is not that their standard of life is not increasing absolutely. It is. It is that it is not increasing fast enough. Their population is rapidly growing, at least as rapidly, I would think, in most cases, as any improvements in their agricultural and industrial productivity. They are urgently and chronically in need of capital for economic development, but I should doubt whether their total requirement, even if it had been fully calculated, was not wildly beyond the capacity of the industrialized countries to provide at any time within the next ten years. Their most salient political characteristic is often a nervous and self-assertive nationalism which is more likely to show resentment at the extent to which Western provision falls short of their own demands, than gratitude at the extent to which their need is actually provided for; and more likely to demand the things—even down to armaments—which bring a return in prestige or propaganda, than those which a scientific appreciation, objectively undertaken, would actually show to be in accordance with the interests of their peoples. Experience tends to show that in the nature of the case we are bound to distribute our assistance widely; our Communist rivals can concentrate theirs where they are likely to achieve most effect or cause most trouble. Yet it is precisely upon this group of nations that, in the absence of a major clash, the limelight is likely to play. And because, after all, what is being played is not a game or a charade, and because after all there is dignity in human suffering and urgency and danger in unrequited human needs, we shall neglect their wants and their difficulties only at our own peril; we will be driven to assist them even though the instability of their internal regimes, and the low standards of life of their people, often appear to make them unwilling to accept responsibilities or acknowledge obligations as a corollary of being accorded rights or even given assistance.

Hence, in our community of the West, our own need of one another in the future will be more and not less. For if we are in the end to win over the rest of the world to our way of thinking—the uncommitted nations as one by one they are enabled to break through to industrial prosperity, and finally even the Communist nations, as they are compelled (as I believe they ultimately will be) to turn over to a consumer-dominated economy—then we must first endeavour to build up our own community into a microcosm of what the world itself could be like if humanity as a whole determined once for all to devote itself to the moral values of our own philosophy of liberty under the law. This means a great extension of the political, cultural, economic and technological relations between our countries, continents and systems. It means the creation of a more solid credit base across our currencies. It means the building of interdependent industries based on multilateral trade overstepping political and tariff boundaries. It means the sponsoring of international dams, bridges, waterways and roadways. It means joint projects of research and development. Nor is this all. Why should we not, if our scientists advise, march into space on multinational vehicles instrumented to carry out joint experiments? Why should we not, as our resources allow, go forward together to clear the forests and speed the plough in lands yet unharvested, and bring power and light, transport and communications, to countries in which human beings have hitherto been beasts of burden? Let our Universities and colleges develop mutual ties and links which will consolidate an international republic of learning—the only aristocracy

through time. We are all the children of Cain, with his mark upon our brow. There is scarcely a nation which does not owe its genesis or its survival—or, more commonly both—to bloody acts of revolution or of war. But the fact that the problem of violence in human affairs has hitherto proved intractable should not necessarily lead us to shrink from its present solution. Many of the simplest devices in human history have been due to discoveries once and for all at a particular juncture of time. All human civilizations, for example, have been conscious of the need for writing. But the alphabetic system of writing was discovered in human history once-for-all in the little Lebanese village of Jebail, the ancient Byblos, and all existing alphabets derive from that single source. All human calculation demands the existence of an arithmetic. But the single discovery upon which all existing systems are based derives from the once-for-all invention of the zero by some anonymous Hindu genius, reaching the West from the Arabs. All modern administration demands a civil service, entered by competitive examination and removed from party politics. Without it democracy and dictatorship alike are hamstrung. All existing civil services derive from the Chinese, brought over to Europe, through India, by the British, and copied from them by every civilized society. Representative Government, Federal States, even modern Cabinets, are all devices to deal with problems previously intractable, but now confronting civilization with no insuperable difficulties. Each has been discovered, or evolved, at the moments and places in the historical process when need and genius confronted one another in a fruitful union. May it not be that our own scientific age, faced with the danger of nuclear annihilation from failure to deal with the age-long problem of violence, will throw up the remedy from the pressing and desperate nature of the need?

What I wish to do is to put forward a partial explanation of the failure of the 'law not war' attempts to date. My judgement is that this failure has been due to attempts to outlaw war without really substituting law, or anything like it, at all. My thesis is that the crying need is for a whole new range of legal conceptions in the international field, starting from the fundamental issue of what are (or rather should be) the juridical persons recognized by international law, and proceeding to a full code of behaviour for civilized states covering the whole range of conduct at its sensitive spots, capable of being ascertained judicially and capable in principle of enforcement—whether or not, in practice, the means of enforcement is found to exist. My claim is that such a code does not exist at present, and that until it does 'law not war' can never be a reality; my claim, further, is that the legal professions of the free nations could perform an indispensable service by disentangling the issues, making proposals for a code, and offering advice as to the means of establishment.

My criticism of existing attempts, from Metternich onwards, is that they tend to rush their objective by prohibiting or limiting self-help before defining substantive rights and obligations, or offering adequate advice about the ways in which there could be established a strictly judicial ascertainment of rights and obligations, at least in principle, founded on reason and justice and a consistent body of legal reasoning. This is not in fact the course which has been followed. Politicians have congregated at international congresses with the express object of winning the last peace. They have sought to prevent the kind of thing which happened last time, by endeavouring to put a stop to a particular kind of appeal to force, without pausing to ask themselves whether they are morally justified in so doing, or even if they have a reasonable prospect of success until they have evolved a body of rules of conduct and a series of legal conceptions which really stand up to the wear and tear of actual international relations. The evolving state of affairs has produced a series of institutions incapable of performing the tasks assigned to them, and a ramshackle corpus of international law founded partly on conceptions which are themselves the cause of repeated recourse to violence, and partly consisting in gaps neatly designed to avoid discussion of the very topics requiring urgent consideration.

Of course this would not matter—or at least would not be difficult to remedy—if there were not a connection between law and justice, and between justice and morality. Legal positivism could easily be made to fit any case once the need for a code were established. But the West has always rejected legal positivism, and habitually returns to the position that there is a justice independent of positive enactment, and a law and morality for human beings 'natural' at least in some sense of that overworked and ill-used term. I will not here argue the philosophical case. I will simply say that I agree that the whole conception of 'law not war' depends in the last resort on the truth of the immemorial Western philosophy of a 'natural law' firmly rooted in morality and justice.

The existing structure of international law, what I might call the Common Law of Nations, has grown up by customs as amended and supplemented from time to time by the contractual nexus of treaties which is, in the absence of a legislature, the nearest thing to statute law, and to a constitution, that the world has got. This somewhat diffuse body of doctrine does take some account, at least in its origin, of this kind of idea. But it will, I imagine, be generally conceded that it grew up in Europe not much later than the seventeenth century, in an era in which the modern conception of sovereignty had superseded the earlier notion of a more or less unified Western Christendom, and from which, in consequence, the theological teaching concerning a just war had been relegated from the world of jurisprudence to that of morals and politics.

The consequences were two-fold. In the first place, international law, and the very documents which have since been designed from time to time to supplement and give effect to it, are largely founded on the legal conception that the world is divided into sovereignties and that the only fully juridical personalities in the world of nations are sovereign states. I wonder whether this is even plausible as a fact, and whether, if it is plausible as a fact, it can be said to give rise to a series of legal conceptions adequate to the purpose which we have in mind. If, as I think, this conception of a self-sufficient sovereignty is actually breaking down in the modern world, the juridical framework to which it gives rise may well turn out to be irrelevant or inappropriate.

Of course, the idea that states are in themselves juridical personalities was a convenient one, and derived by an easy analogy from the law of corporations, made easier by the historical fact that sovereigns were originally human persons at the head of states. But many of the controversies which lead to the outbreak of violence have had to do with the formation, dissolution, amalgamation, or territorial delimitation of states. And it is here, of course, that the analogy with corporations or human persons is least convincing. States are not things given by nature, like natural human beings; nor are they entirely artificial like corporations. Moreover, the minorities of human beings within states are sometimes natural entities which have at least a superficial resemblance to the same sort of community as those which, in a separate geographical area, have gone to make a state; and in some cases the existence of such minority communities—in Czech-

87

oslovakia, Poland and Yugoslavia, for example—has been of the very stuff of which international conflict is made.

The same degree of unreality dogs our footsteps when we create international assemblies composed of states. The relationship between New Zealand, for example, and the United States, or between, say, Iceland and India, is in no relevant respect analogous to the difference between a big man and a little one, or a wealthy corporation and a poor one. In the latter case the relationship is that of a group of a few hundreds of thousands of human beings to one composed of hundreds of millions. Nor is it easy to adopt a single criterion alternative to one-for-one equivalence between states, such as population, wealth, or influence.

Or again, to what extent can the internal affairs of states be ignored? The Charter of the United Nations says clearly that they should be. This is in line with the classical doctrine; and indeed we cannot forget our experience of Nazi Germany and other powers who have sought to raise a supposed right to champion the alleged grievances of minorities within a state as an excuse for aggression. (This, incidentally, was one of the international habits of Czarist Russia which gave rise to controversy.) But to what extent can they be ignored? What are we to say, for instance, about the present régime in Hungary, or even in South Africa? Is the underlying conception acceptable? Can a system of international justice which is to be the foundation of 'law not war' be founded on a recognition of sovereign states as the only legal personalities? Personally I would doubt it. But what other more acceptable principle is there—and if we abandon principle, what exactly becomes of our campaign for 'law not war'? If law be not founded on ascertainable principle, wherein lies its foundation?

But, secondly, the foundation of international law upon the

classical doctrine of sovereignty has inhibited the growth of a juridically determinable code of behaviour covering the really sensitive points of national conduct. Yet it is surely useless to talk of 'law not war' unless a code of acceptable rules does precisely this. This is one reason why the World Court cannot of itself solve the problem. By far the greater number of issues cannot be referred at all because, owing to gaps in the rules, they are simply not justiciable. Of those that are justiciable, the jurisdiction is often unacceptable to one side, since in the absence of a method for altering the law the World Court is driven virtually to defend the *status quo* which may be obsolete or unjust. It is notorious that the classical international lawyers rejected the idea of a just war as something capable of legel definition. Apart from other difficulties, to do so would have contradicted the idea of sovereignty. The consequence was, there were two separate sets of rules, governing respectively the conduct prescribed for nations at peace, i.e. those who had no dispute between one another, and the conduct prescribed for nations at war, i.e. those who had already chosen to resort to self-help to achieve what they claimed was justice. But it had relatively little to say as to what was justice at the point at which peace was likely to turn into war. And, although numerous treaties have prohibited or limited the appeal to self-help, none of them go the whole way to give a code of rules for regulating conduct which could define justice at any given state of time, and so do away with self-help as a means of attaining it. I would think, myself, that such a code is absolutely necessary before 'law not war' can be a reality. Unlike some people I do not myself think that it is in principle impossible of attainment, and I would think it very much the business of the legal profession to strive to attain it.

In two respects, recent efforts to get rid of self-help in

international affairs have actually tended to make matters worse. Under the influence of two world wars precipitated by aggressive invasion, these attempts have tended to limit self-help to resistance to aggression of this kind. This may or may not have been wise, and certainly cannot be written off as unwise without careful consideration. But there are many wrongs—political, economic, and even military—which one nation can inflict on another in such a way as to make life intolerable or difficult but which do not consist in, or involve, territorial aggression. Granted that self-help is undesirable in such a case as inappropriate, excessive, or as tending to favour the strong power, the mere removal of self-help as a remedy, without formulating rights and providing an alternative means of giving effect to them, simply puts a premium on lawlessness in all its forms short of physical invasion. In an age of confiscation, and of aggression by subversion or infiltration, this is hardly good enough, and, although it may be said that some of these forms of aggression, such as subversion, have been explicitly condemned by the United Nations, they have never been precisely or effectively defined, and methods of enforcement have not been found.

The attempt to limit self-help to actual resistance to invasion has led to one further unfortunate consequence. A carefully prepared all-out assault can in theory be mounted in such a way that, under existing procedures, the victim might have to await his fate with docility, while other powers were unable to help, in which case he might be annihilated before he was able to appeal for assistance. In this context I cannot omit to observe that had not the United States and Britain entered Lebanon and Jordan respectively in 1958, by invitation of the local authorities, the conclusion seems irresistible that, by one means or another, two whole nations might have been obliterated almost overnight. But the status of licensees which both countries obtained and used as the juridical basis of their action on that occasion could hardly be relied on to exist in every case. In some cases—notably that of Hungary— the character of the local authority was such that the invitation was itself, in the eyes of most of us, an added provocation and an added contempt of justice and morality.

I am sorry to have painted so gloomy a picture. But it is at least worthwhile realizing the extent of the difficulties in front of the 'law not war' movement, and that these difficulties are deeply rooted in the well-established characteristics of international law. I have now a practical proposal. I recognize that many of the problems I have posed can be solved, if at all, only on the political front. Yet I would think that all of them would gain enormously by being studied objectively as legal problems, and specifically as projects for research, by a body of non-political lawyers of the free world—that is, lawyers sponsored by their professions or by foundations and not by their Governments, and lawyers sponsored by professions, such as the English and the American, devoted to the conception of liberty under the law as the philosophy of human relationships which it is their desire to promote.

The terms of reference of such a body would have to be of the widest possible. For my part, I would hope that three broad lines of investigation might reasonably be established. I would hope, first, that an investigation would be made into the desiderata technically to be expected of a system of law which would take the place of war. I would hope that this investigation would be made without necessary regard to the present established doctrines. Given the authority to legislate a just system, what sort of system would we, as professional lawyers, set out to create? Secondly, I would hope that we could examine the deficiencies, real or alleged, of the existing system. What are the true gaps in the present code of conduct? Can we devise a system of rules for the behaviour of nations towards one another which would make the main breaches truly justiciable, and not simply raw material for diplomacy? Could we outline, at least, types of conduct falling short of territorial aggression which could be condemned, and can we think of methods of preventing them? In other words could we take up the question of sanctions where the League of Nations left off? Thirdly, can we make juridicial sense of an appeal by minorities, or by other powers, to make some—and if so what—'internal' matters the subject of judicial inquiry? Some attempt has been made to achieve this by the Human Rights Convention to which some European states, including Great Britain, have adhered. Has it worked? To what extent could it be universalized? And what juridical persons ought to be recognized?

The essence of such an investigation would, in my view, be that it should be unofficial, that it should be limited in the first instance to lawyers, and (so as not to dim the prospect of constructive proposals) that no profession should be entitled to be represented unless belonging to a country subscribing, at least in theory, to some body of jurisprudence based on liberty under the law. At a later stage I would like to see the adoption, by our own and other like-minded countries, of the code of conduct which I hope would emerge from such a study as the legal basis for the kind of interdependent community which I sketched earlier.

*　　　*　　　*

It would obviously not be possible to include the Iron Curtain countries in such a plan, at any rate at its inception, until and unless the political climate becomes more favourable than it seems to be at present. But in general it is implicit in the creation of an interdependent society that should our separated brethren on the other side of the Curtain become minded to join us, in any or all of the enterprises involved, they should be welcome to co-operate, though never to dominate or obstruct.

Meanwhile it is still our interest and our duty if possible to reduce political tension in the world, to leash so far as we can the destructive powers that have been put into the hands of man since the splitting of the atom, and to get some agreement which will make it possible to do away with the testing of nuclear weapons and, ultimately, and as part of a general programme of disarmament, get rid of such weapons themselves. It is not that I would do so unilaterally. It is not that we should do so without safeguards. Most people see that there must be inspection and control and that, in the main, it would be foolhardy to move without them. It is not that, if no agreement proved possible, I would have any doubt that even the dangers of the present uneasy equilibrium are preferable to the abandonment of essential principle. It is that the whole nature of the so-called deterrent has altered in the period since the end of the war. Once, it was a deterrent in the true sense, since we were deterring a conventional attack by the possession of an armament which only we possessed. Now, we are relying on mutual terror in a field of weapons possessed by both; and, though it is true that such a balance of mutual terror is efficient now, and may be efficient for years to come, yet, viewing the matter, as I endeavour to do, against the backdrop of history, I am solemnly convinced that if we go on indefinitely experimenting with these weapons, manufacturing them and stockpiling them, boasting of their potentialities, and keeping them at instant readiness, sooner or later a situation will arise,

93

sometime, somewhere, where one will go off. If it does, it will give rise to a chain reaction not less predictable because its course and causes are in the realm of politics and not of physics alone. It may be the deliberate explosion of a sophisticated weapon by one of the major powers. It may be a misunderstanding, or the blunder of a trigger-happy subordinate. It might be the crude explosion of some clumsy and primitive nuclear device by some adventurer in the second or third rank of world powers who had clandestinely acquired and manufactured it for reasons of prestige. It might be fifty years from now—or it might be ten, or it might be two. But I have children at school, and I would wish that they might live to a ripe old age, and that we might bequeath them a state of the world which will permit them to do so.

There is a real sense in which this may be said to be the biggest and the most urgent of all the questions thrown up by the inter-action of science and politics. I recognize the risks, but in taking them prudently and together I believe we would be securing a future in which there was some hope for humanity in exchange for one in which there is ultimately none. If by our mutual arrangements we are sufficiently protected against treachery or surprise, I am not afraid of the challenge of co-existence—provided only that we face it united.

Chapter 5

THE RELIGIOUS BASIS

————————————————★————————————————

The Reformation destroyed the unity of Western Christendom and overthrew the teaching authority of the Church on which it had been based. The scientific discoveries and hypotheses of Lyell and Darwin shattered the kind of interpretation of the Bible which had been built up in Protestant Christendom to take the place of the teaching Church. Free discussion and popular education after a long period of ignorance would, I am inclined to think, have led naturally, and almost healthily, to a period of doubt and self-questioning, even if the other factors making for infidelity had not been operating. But the movement took a definite form. There is no doubt that of the various political revolutions which have taken place since the turn of the century the majority have been consciously anti-Christian. In the democracies where the same political forces are at work in a free society the movement has taken the form of the worship of Mammon and Venus; in the dictatorships that of Moloch or Mars. In the unthinking multitude these movements have succeeded to the extent of creating indifference; in organized minorities they have actually produced active hostility to the Church.

This movement of human thought all over the world, towards a materialist or positivist view of the Universe and away from a religious or even an idealist philosophy, has coincided with a real and very obvious retrogression from the humanism (in the sense in which Erasmus used that word) of the nineteenth century towards the abominations of Belsen and Buchenwald, the horrors of Hiroshima and Nagasaki, the rape of Budapest and all the cruelties and inhumanities of our present age. More men and women have been tortured, more done to death more brutally, more starved, more mutilated, more persecuted, more condemned to rot their souls in prison in the so-called century of the Common Man than at any time since the Dark Ages.

To me at least there is a close connection between these two main movements of thought and practice—the conscious abandonment of religion and the idea of God, and the retrogression from humanity and the nineteenth century softening of manners. They are, I believe, related to one another as cause to effect, and they are at the root of nearly all the troubles with which the world order is face at the present time. I cannot, therefore, conclude these personal reflections upon the inter-action of science and politics without discussing the need for faith in a scientific age. Given the validity, and understanding the nature, of scientific hypothesis and reasoning, what can a man intelligently believe? Does it matter? Will anything do? Or is the scientific method enough—that is, should a man conscientiously refuse to believe anything whatever which cannot be logically incorporated in the general body of scientific reasoning? These are all questions which everyone of a certain intelligence must have asked himself. The quality of our society, and quite possibly its continuance, depend upon the answers we feel able to give.

I begin by asserting my passionate belief in the importance of sound general ideas. They do matter. It has been the terrible experience of the last fifty years that they matter more than anything else. I have come to believe that Plato was correct in teaching that there is no surer recipe for anarchy or tyranny in society than the absence of sound general ideas, or the possession of some half-baked but emotively powerful 'philosophy', ideology or creed. Better nothing rather than that; unfortunately nature abhors a vacuum, and however desirable it may be, it is, I am sure, intellectually impossible to believe nothing.

Because of this, my plan is to argue the case for the reasonableness, the intellectual respectability and validity—more, the practical necessity—for metaphysical speculation; to say a word about its status in relation to science and demonstration; and to give some account of the nature of my intellectual journey along its paths since I read philosophy at Oxford over thirty years ago. I do so with humility and diffidence, because I am aware that my long immersion in practical affairs has rusted my philosophical equipment and made me out of sympathy with much that appears to find favour with the present school of philosophers. Yet I also write with conviction, because, unlike many people, I am convinced that this is an intellectual age, and the present generation realizes, perhaps more than any other, that it is no good praising a philosophy for its utility or beauty unless one has first convinced oneself of its reasonableness and truth.

*   *   *

Religion, philosophy, and science all start from the same point—an unconquerable belief in the intelligibility and rationality of the Universe. None of these studies would be possible unless this was the real driving force behind them. All, moreover, really produce results deriving from the same

G   97

mental and moral qualities, and occupying fundamentally the same status. In the inner recesses of their understanding, theologians, philosophers and scientists all know that their most confident assertions contain an element of the hypothetical, and achieve at best a high degree of probability. Nothing is gained, and much is lost, in claiming more for any scientific theory, philosophical reasoning or religious dogma. All depend for their discoveries as much on intuition and poetic imagination as on careful reasoning. All can be overthrown in the end by careful reasoning whatever the elegance or beauty in the imagined hypothesis. We live without certainty in the discipline of a darkness which is still not wholly without illumination.

Yet in the nature of belief itself, embedded in the logic of the very concept, there is something that tells us that absolute truth and absolute certainty are not mere myths. The relative and the contingent may be all that we can ever have. But relative and contingent are words which only have a meaning in a world where the absolute, also, reigns. Of course we cannot, whether by science or philosophy, think ourselves out of the limitations of space and time, or outside the experience of our flesh. These are not only built into our experience—if that were all, we could then discount them— but also into the means of our experiencing. Yet to deny that they are limitations, and that there is, outside and beyond them, a reality more absolute than either, is to involve ourselves in contradictions and difficulties as wild and insoluble as any which may be created by whatever assertions we care to make as to the kind of thing that such a reality can be. Even Russell's 'set of all sets' sits on the top of his intellectual palm tree, grimacing and pelting him with coconuts as maliciously as Kant's *Dinge an sich* or the ἄπέζαντν which baffled the Greek philosophers who set up a πέζας. The more we deny the existence of an absolute, the more confidently we are involved in the assertion of it. So neither the scientist, nor perhaps even the lawyer, can complain if the poets, the philosophers, the theologians, and even the mathematicians, have dreamed and speculated about a world outside that into which we are integrated.

Apart from its practical glories, I should have thought that one of the achievements of modern science is that, on a purely intellectual plane, yet on purely empirical foundations, it has killed materialist philosophy stone dead. To Locke, the world of sense data in a Euclidean space flowing through a well-explored dimension of time made pretty good materialist sense, although, inevitably as we now see it, the ultimate irrationality of these ideas led inexorably to the philosophical agnosticism of Hume. Yet, even when the subjective character of colour, taste and sound had been exposed, there was a sufficient reality in the so-called primary qualities and in the categories, or dimensions, of space and time, to give us a more than sneaking sympathy with Dr. Johnson when he kicked the stone and disposed of the doctrine of Berkeley with the contemptuous, 'Sir, I refute it thus'.

Yet, whatever discipline we follow in modern science, Johnson's stone, and for that matter Johnson's foot which kicked it, have disappeared—if not into nothingness, at least into something quite different from anything which Locke could have imagined. From the point of view of the cosmic particle, approaching Johnson's stone for the first time from outer space, the stone presents more interstices than solid obstacles. But the solid obstacles themselves, if at that level they can be called solid, are really mathematical concepts. Are they waves? Are they charges of electricity? Are they tiny particles? They are all, and none. It depends how, and why, you look at them. They have neither colour, taste, nor sound. Yet, they have mass and dimensions—but mass and dimensions themselves are contingent and qualified hypotheses. The particles are mathematical conceptions which can only be appreciated as the result of the most formidable body of calculations and inferences, and even then can be formulated only in terms of hypothesis, however probable. Whatever they are, they have qualities quite different from the material conceptions of the past. They were learned of by scientists in whom observation and deduction were closely married to a creative insight.

In the meantime it is worth casting a glance at Johnson's foot. This time, let us put on the spectacles of the organic chemist. From his point of view, Johnson's foot is a very much more majestic structure than was the stone which we imagined through the eyes of the cosmic particle. Johnson's foot appears to the chemist as a mass of macro-molecules in the form of intertwined spiral staircases of combined oxygen, hydrogen and carbon atoms, so complicated that each requires a model too huge to be assembled physically in the laboratory. Even that, however, is only the beginning of the story. The stone and the foot are simplicity themselves compared with the whole organism which we call Dr Johnson.

For it must now be pointed out that consciousness itself is a phenomenon, no doubt capable of being studied scientifically, but nonetheless of a nature which, on purely logical grounds, excludes a purely materialist conception of any Universe which contains it. However it is produced, whatever the chemistry, the physics, or for that matter the electronics of the brain, the thing which says in regard to the reality it studies, including itself or its fellows: 'I know, I believe, I deny, I understand', or, worse still, 'I love, I admire, I detest', is not a thing which can be described simply in terms of molecules, hormones, ionized particles, or electrical activity—unless, of course, as has been seriously suggested by philosophising scientists, these things in their turn have a 'within-ness', an inner character of life depriving them of the right to be considered solely in their mechanical or material aspects. But, if the latter is the case, it is in consciousness that this 'within-ness' becomes critical and apparent, and it is consciousness itself which forms at once the crucial case for the study of 'within-ness', and the most convenient field for observing and reflecting about its characteristics. In a sense, the conscious being is the apex of the evolutionary pyramid, and though much can be learned by delving about at the base, it is the conscious being itself, the most complex product of the evolutionary process, which raises the most challenging questions about the nature of reality, and whose ultimate destiny and future confront us with the most intriguing mysteries to be solved.

I say destiny and future. But this makes necessary a further observation about the world which takes us even further from the materialist view. Whatever may be said about space, time—so far as we can tell—is in its innermost being directional. This is why it is not merely a dimension. Whether we consider the law of entropy in the physical world, or the doctrine of evolution in biology, or even the more limited field of human psychology, time flows in one direction only. It may flow, as the hymn says, like an ever-rolling stream. But it never flows backwards. It is as true now as when Agathon wrote the verse for the Greek Theatre:

> *One thing at least God never knew,*
> *How, what is done, once done, to undo.*

There are some people who argue whether these doctrines of science have not made impossible the traditional and orthodox tenets of religion. This I shall examine in due course. What needs to be said now is that such doctrines have dissolved materialism into mist. Professor Lovell writes, in effect, that there are two conceivable cosmogonies for the scientist—the doctrine of continuous creation à la Hoyle, and the doctrine of the explosion of a primeval atom. It may be that neither involves—though I would have said neither can conceivably exclude—a purposive creation, once-for-all or continuous, by an intelligent Deity. What is clear, however, is that both demand something more than a constantly changing, self-adjusting, and entirely self-explanatory Universe. Both involve creation of a kind, even if we are careful to add that we do not wish to beg any questions about a Creator. If nature has not an end, it at least has an unknown origin, and a direction, whichever of the two views is right. Neither makes the Universe like a perpetually self-winding watch. Something has to be added either at the beginning, or all the time; and as for time, it flows like Ole Man River only one way.

But whatever may be said about the plausibility of a purposive Creator, the human mind is as intrinsically purposive in character as time is intrinsically directional. Indeed, the two facts are closely connected. No doubt the scientists can tell us a great deal about the origin of the value judgements, and, if we include the systematic studies called, or mis-called, 'human sciences', perhaps even about their character. No doubt the doctrine of evolution can tell us a very great deal about the origin, and the survival value, of our moral, political, and—within limits—our aesthetic judgements though I have often wondered what survival value, if any, attaches to our exclamations of joy at the sight of a beautiful landscape. No doubt also, as in the field of general consciousness and cognition, there are useful analogies from the animal kingdom. But, at the end of the day, at the apex of the evolutionary pyramid, we are faced with the spectacle and the problem of a conscious being—or rather a race of conscious beings—who say of the reality they study, and of which they form a part, not only 'I know', but 'I will'; not only 'this is true', but 'this is good, or just, or ugly'.

Moreover, the fact of consciousness and its purposive nature has, as Professor Medawar pointed out some time ago, developed a new meaning for genetics and human evolution themselves. When the Christian Church proclaims itself the new and the true Israel, it does not mean that the genes and chromosomes which produce Christian bodies and brains are physically connected with those of Abraham. It means, and Professor Medawar meant, that ideas and ideals, passed on by writing and oral tradition, communicated by symbols and words, can produce amongst conscious beings, themselves in part the product of purely physical processes, a new race of men who, in turn, can develop, by an evolutionary process, ever-fresh characteristics, and pass these on to posterity by conscious acts of transmission. Evolution has produced a creature apparently capable of evolving himself, and coupled with this portentous fact is the fact that the further he evolves the more completely does the conscious process appear to take the place of natural selection. What kind of a creature is this, and what sort of a Universe is it that gives him birth?

In the process of conscious evolution, the human person is not only conscious and purposive. He may be, and often is, disinterested. He is capable of self-sacrifice, and of love. We must be careful not to make too much of this. No doubt evolutionary science has some useful things to say about the origin and value of this phenomenon. What survives natural selection is not an individual, but a group, a constantly palpitating and altering colonization of living entities. If you care to call this 'a species' or 'the human race', you may do so, though in doing so you will miss part of the reality. For both expressions involve the notion of something fixed, and set, and complete, and not constantly moving, altering, pulsing, contracting, and expanding. What has survival value has accordingly value not for the individual but, if you will, for the species, even though, as a humanist would go on to say, the species is exemplified, and therefore presumably possesses its value, only through and in the individual. Survival value therefore must be related to the survival of a group rather than an individual, and performs its biological function if it achieves that survival even at the expense of the individual's own existence. Biologists are surely on unassailable ground in finding an evolutionary origin and value in the psychological instincts which make us good citizens, good parents, or good Christians.

But this, of course, does not in itself do justice to my present point. For, once again, at the apex of the evolutionary tree, has emerged a creature who can consciously reflect about, and adopt, or indeed reject, what I have called love as the central theme of his individual life. The fact that, like hunger, sex, or other natural instincts, it possesses a strong survival value in the purely evolutionary sense, does not deprive it of its interest to us as, in the human being, a conscious activity forming an intrinsic part in human purposiveness. After all, human beings have to come to some sort of terms with all their other natural instincts. They can, and do, control and regulate eating, mating, and other behaviours. There would seem to be no reason whatever why human beings, after reflecting about the complex of feelings and instincts out of which what we call love has emerged, should not deliberately suppress, atrophy, or limit them. And indeed this is precisely what has happened. In our own time, we have seen the feelings of loyalty, social solidarity and unselfishness used to bolster and support precisely those things which are most detestable or ridiculous: race, class, or national consciousness carried to the point at which they become in substance conspiracies against the rest of the human race—individually because they deprive the individual of any semblance of dignity and decency, collectively because they establish relationships between groups in which one simply subserves the interests and purposes of others. Thus disinterestedness itself is as much capable of perversion as any other human feeling.

The real philosophical challenge, however, is presented not by the perversion but by the flowering of Man's qualities. No doubt error presents us all with philosophical problems in the field of cognition. But the real question is not why the reasoning powers of a materially evolved being are capable of error, but what sort of a Universe it is in which a materially evolved being can by his own 'within-ness' arrive at even an approximation to truth. Truth, not error, is the real philosophical problem; goodness, not evil; light, not darkness; love, not its perversions. For in the purposive process by which the human species has deliberately evolved itself from time to time, love—that is the capacity and desire for individual self-sacrifice—has acquired an entire set of ideas of its own. No civilization can exist without laws, and the basis of any acceptable code of laws is the belief in a pre-existing order of rights and responsibilities—which I might refer to collectively as justice—which have an inherent authority of their own not entirely derived from the political authority of the legislating body. Indeed, in the history and evolution of law, the idea of a legislating body is a comparatively late development, and its exact functions and power are the subject of considerable controversy.

It would, of course, be highly desirable for me to expand the argument in a variety of different ways. But I have said enough to show why, for purely intellectual reasons, which have grown stronger as I have grown older, I feel compelled to assert that the natural order itself is not such as to be susceptible of a materialistic explanation. Whether we look to physics, and postulate a continuous creation of hydrogen atoms in a continually expanding universe still continuously subject to the second law of thermo-dynamics; whether we look to biology and examine man in his true context in the evolutionary chain of living species; whether we embark on the more speculative fields of mathematics, philosophy, or even theology—we are never able to regard the natural order as self-explanatory without involving ourselves in a network of contradictions. Nor are we able to regard a world in which we find creatures like ourselves existing as one completely materialistic in character. In short, no one who examines the world in a complete spirit of scepticism can afford in the end to ignore religion—that is, a view of the world altogether different from the mechanistic; one in which matter, if there is any meaning in this term, and mind or spirit, are at least co-ordinate and equal realities, closely inter-related.

*　　　*　　　*

What religion is, and whether one should choose one, are of course very different matters. There would be something to be said, from the premises of the argument I have been outlining, for a kind of pantheism—an inner within-ness of everything emerging from a state of *Urdummheit* or primeval stupidity and struggling to express itself in ever higher states of consciousness, the human species being the highest so far evolved, but perhaps offering, through its collective consciousness, an opportunity for further advancement. An odd world, perhaps, but more rational, I would think, than a purely materialist conception of the Universe. I suppose that quite a superior morality could be evolved from a system of this kind, though I cannot for the life of me see why anyone, though it were evolved, should adopt it as his own.

But the question is: what sort of a thing is a metaphysic, and is it possible—or desirable—to have one? I suppose most philosophers and scientists today put the possession of metaphysical beliefs outside the bounds of the rational, perhaps even of the intelligible. There is a sense in which I agree with them. If by the possession of a metaphysic is meant either the ability to infer from fairly every-day materials a watertight theory of the construction, purpose and destiny of the Universe, or thereafter the ability to demonstrate it to the conviction of one's fellow-men, well then, I must say I regard the whole project as absurd, and possibly the conception as meaningless. But what I look for is something less ambitious and, in my view, more satisfying. I do not expect certainty in this life about most of the things which really matter. I want a working hypothesis that satisfies me intellectually and purposively, that gives me direction, hope and objectives, and still seems to me to explain the main facts of the world as I see it.

This much must be conceded. Whatever may be ultimately true or right, the main nastinesses of our time have offered exactly that—and still do, for that matter. But can an educated man not hope reasonably for the same sort of thing—stipulating only that it may be true and not false, and honourable and not nasty? I cannot rid myself of the belief that the need for such a working hypothesis is deeply rooted in human nature itself—and, as it seems to me, legitimately rooted, if the scientist and the theologian are right in their belief that the world is an intelligible place and that at the end of the day there is sense to be found in it if we know where and how to look.

It is at this stage that the evolutionary pantheism which I have endeavoured to describe, not I hope wholly without sympathy, seems to me to break down utterly. A world in which something which does not yet exist is striving dumbly to bring itself into being by means of an evolutionary process whose only weapon is natural selection, but which it is slowly evolving itself into a position to control, seems to me to make no kind of sense at all. If it did make any sense, I do not believe that a man could rationally identify himself with something like this with any sense of purpose or dedication. If consciousness, if purposiveness, if morality, and love and self-sacrifice and justice, are, as I have tried to demonstrate that they must be, at the end of the chain of evolutionary phenomena, it can only be, it seems to me, because intelligence, purposiveness, yes and justice, love and self-sacrifice too, are themselves at the centre of reality, not in an inchoate but in a transcendent mood, not as abstract qualities like the Platonic ideas, but because they inhere as such qualities only can in transcendent personalities. In other words, I claim that the only viable construction of the complicated facts of physical and biological science and psychology, the only true reading of human experience, remains at the end of the day a Theistic one. I say it with humility because I realize how intellectually unfashionable at the moment such a view must be. But, having arrived at it, I must quite frankly set out some of the inescapable conclusions which seem to me to flow from it.

The first is that what starts as an intellectual conviction becomes by its inherent logic an overwhelming passion, a devouring practical necessity to embrace and identify oneself with the ultimate reality of consciousness, personality, intelligence and purpose, once this is perceived and identified as a real object of knowledge and desire.

The second is, I would say, that it is only some kind of metaphysic such as this which renders the practical humanity —and, in its original Erasmian connotation, the practical humanism—of which it seems to me the world stands in such need today, a really going proposition, with a motivation and intellectual integrity to make it work.

The third is that at the bottom of the well, at the heart of the matter, at the core of every intellectual, moral, or practical problem, is neither self-seeking nor the so-called—but to me singularly unconvincing—ideal of service, but adoration, surrender, worship, self-identification, and self-sacrifice; and life can be said to be truly lived and truly understood exactly in proportion as this is realized and achieved, and to be trivial, meaningless, or plainly miserable and calamitous, exactly in proportion as we are prevented by circumstance or choice from getting there.

The fourth, and it is to me the most powerful emotively of all, is that when we live, as I believe we all want to, honourably and nobly and well, somewhere at the very centre of reality there is that which responds, not merely naturally but of its own intelligent being and nature personally, to that which we are doing and endeavouring and giving.

This belief, I should have thought, is worth trying as a working hypothesis, and in my view, at least, if this is not worth trying nothing else really is. Tried, it becomes I believe more than a working hypothesis. But at this stage faith, in which we all must live in one form or another, moves from being the substance of things hoped for to being the evidence of things unseen.

THE RELIGIOUS BASIS

A while ago a television interviewer asked me if I found my religion useful. Rather primly, I fear, but none the less correctly, I replied that the question should be not whether it was useful but whether it was true. I now say that, if true, then because true, useful it certainly is. For when I survey the emotional, the intellectual, the moral, the political, even the physical, litter and chaos of the world today, in which truth has almost ceased to be regarded as objective, where kindness is made to depend on political, class or racial affiliations, and only the obvious stands in need of publicity, the conviction comes to me with redoubled certainty that the traditions nurtured and fostered in Greece, Rome and Jerusalem, which I hold sacred, can alone sanctify and therefore civilize the restless and destructive spirit of our time which, allowed to develop uncontrolled, could well annihilate the human species upon the planet, or alternatively undo the process of evolution itself and plunge mankind back into the animal status from which we arose. *Nisi Dominus custodierit civitatem frustra vigilat qui custodit eam.*